wild cats
of the world

Luke Hunter

Illustrated by
Priscilla Barrett

BLOOMSBURY

LONDON · NEW DELHI · NEW YORK · SYDNEY

wild cats
of the world

Luke Hunter

Illustrated by
Priscilla Barrett

Bloomsbury Natural History

An imprint of Bloomsbury Publishing Plc

50 Bedford Square   1385 Broadway

London   New York

WC1B 3DP   NY 10018

UK   USA

www.bloomsbury.com

First published 2015

© Luke Hunter, 2015

Illustrations © Priscilla Barrett, 2015

British Library Cataloguing-in-Publication Data

A catalogue record for this book is available from the British Library.

Library of Congress Cataloguing-in-Publication data has been applied for.

ISBN: PB: 978-1-4729-1219-0

ePDF: 978-1-4729-2285-4

ePub: 978-1-4729-1220-6

2 4 6 8 10 9 7 5 3 1

Maps by Lisanne Petracca/Pantherus

Skull illustrations by Sally Maclarty

Design in UK by Nicola Liddiard at Nimbus Design

Printed and bound in China by C&C Offset Printing Co., Ltd

To find out more about our authors and books visit www.bloomsbury.com. Here you will find extracts, author interviews, details of forthcoming events and the option to sign up for our newsletters.

contents

The cat family

With a global population that may exceed a billion, there are perhaps 300,000 housecats for every Tiger left on earth. The most generous population estimates for all wild cat species *combined* might reach one per cent of housecat numbers; 10 million wild cats, most of them small-bodied, wide-ranging generalists such as Bobcats and Leopard Cats.

The domestic cat is one of the most successful mammals on earth and its most successful carnivore[1]. Resident cat populations occur on every major continent except Antarctica and on most of the world's offshore islands. Cats can survive in virtually any habitat from the Sahara Desert to sub-Antarctic islands, whether they are cared for by people or not. At least half a billion cats are kept as pets around the world, and there are many hundreds of millions more that live as strays loosely associated with humans or completely feral with no reliance on people at all.

The cat's success embodies the evolutionary triumph of the Family Felidae. Felids have walked the earth for around 30 million years and prior to very recent anthropogenic impacts, have been extremely successful. Felid evolution began in Eurasia where the Family's first unambiguous representative – sufficiently different from earlier fossil carnivores to be considered a true cat – is *Proailurus lemanensis*. The oldest *Proailurus* fossils are 25–30 million years old from what was then a vast subtropical forested landscape and is now Saint-Gérand-le-Puy in France. *Proailurus lemanensis* is the likely progenitor of all cat species, living and extinct, that have ever lived. By approximately 18–20 million years ago, *Proailurus* had diverged into two distinct genera that seeded the two main branches of cat evolution. One of these,

the genus *Pseudaelurus* included cats which for the first time in felid evolution had reached the size of the modern Leopard. Their skulls and teeth also carried incipient sabretooth features such that *Pseudaelurus* is now considered ancestral to the felid subfamily Machairodontinae, the sabretooths. This spectacular experiment in felid evolution produced many dozens of species with famously elongated canine teeth and a raft of other modifications in the skull and skeleton that differentiates them from other felids. The sabretooth cats prospered in Eurasia, Africa and the Americas until very recently. The best-known genus *Smilodon* lived until 10,000 years ago in North and South America, and included some of the most extraordinary, largest felids to have ever evolved. *Smilodon fatalis* – the celebrated Californian sabretooth known from over 1,200 specimens in the Rancho la Brea tarpits – was as tall as the modern Tiger but was more heavily built and weighed more, while the massive South American *Smilodon populator* far out-weighed any living cat at close to 400 kilograms. Both species lived alongside humans.

In parallel to the great proliferation of sabretooths, the second major branch of felid evolution arising from *Proailurus* took shape as the subfamily Felinae, the conical-toothed cats. The Felinae began with *Styriofelis*[2], a genus of relatively

---

1 *This book uses the term carnivore as used in scientific nomenclature, i.e. only to indicate species of the Order Carnivora.*

2 Styriofelis and Pseudaelurus *are presumably closely related genera. Earlier classifications recognise only* Pseudaelurus *and assume that various* Pseudaelurus *species gave rise to the two major felid evolutionary branches.*

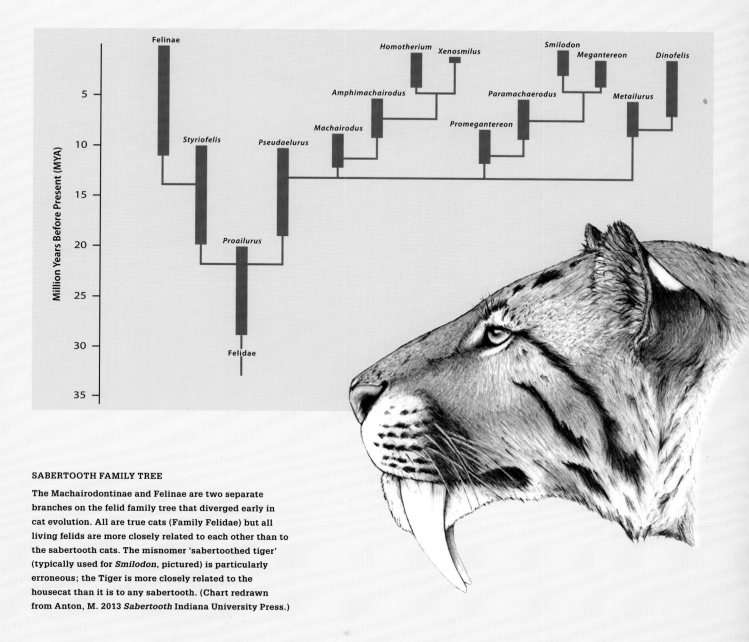

SABERTOOTH FAMILY TREE

The Machairodontinae and Felinae are two separate branches on the felid family tree that diverged early in cat evolution. All are true cats (Family Felidae) but all living felids are more closely related to each other than to the sabertooth cats. The misnomer 'sabertoothed tiger' (typically used for *Smilodon*, pictured) is particularly erroneous; the Tiger is more closely related to the housecat than it is to any sabertooth. (Chart redrawn from Anton, M. 2013 *Sabertooth* Indiana University Press.)

Conversions

Throughout this book measurements, weights and areas have been provided using the metric system, however, those more used to the Imperial system may find the table on page 236 useful.

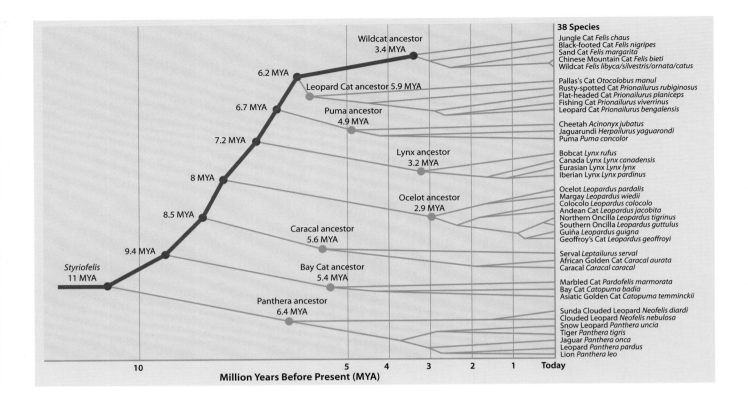

38 Species

Jungle Cat *Felis chaus*
Black-footed Cat *Felis nigripes*
Sand Cat *Felis margarita*
Chinese Mountain Cat *Felis bieti*
Wildcat *Felis libyca/silvestris/ornata/catus*

Pallas's Cat *Otocolobus manul*
Rusty-spotted Cat *Prionailurus rubiginosus*
Flat-headed Cat *Prionailurus planiceps*
Fishing Cat *Prionailurus viverrinus*
Leopard Cat *Prionailurus bengalensis*

Cheetah *Acinonyx jubatus*
Jaguarundi *Herpailurus yaguarondi*
Puma *Puma concolor*

Bobcat *Lynx rufus*
Canada Lynx *Lynx canadensis*
Eurasian Lynx *Lynx lynx*
Iberian Lynx *Lynx pardinus*

Ocelot *Leopardus pardalis*
Margay *Leopardus wiedii*
Colocolo *Leopardus colocolo*
Andean Cat *Leopardus jacobita*
Northern Oncilla *Leopardus tigrinus*
Southern Oncilla *Leopardus guttulus*
Guiña *Leopardus guigna*
Geoffroy's Cat *Leopardus geoffroyi*

Serval *Leptailurus serval*
African Golden Cat *Caracal aurata*
Caracal *Caracal caracal*

Marbled Cat *Pardofelis marmorata*
Bay Cat *Catopuma badia*
Asiatic Golden Cat *Catopuma temminckii*

Sunda Clouded Leopard *Neofelis diardi*
Clouded Leopard *Neofelis nebulosa*
Snow Leopard *Panthera uncia*
Tiger *Panthera tigris*
Jaguar *Panthera onca*
Leopard *Panthera pardus*
Lion *Panthera leo*

Wildcat ancestor 3.4 MYA
6.2 MYA
Leopard Cat ancestor 5.9 MYA
6.7 MYA
Puma ancestor 4.9 MYA
7.2 MYA
Lynx ancestor 3.2 MYA
8 MYA
Ocelot ancestor 2.9 MYA
8.5 MYA
Caracal ancestor 5.6 MYA
9.4 MYA
Bay Cat ancestor 5.4 MYA
Styriofelis 11 MYA
Panthera ancestor 6.4 MYA

Million Years Before Present (MYA)

10   5   4   3   2   1   Today

FELIDAE FAMILY TREE

Comparing the DNA sequences of specific genes in every living cat shows their inter-relatedness; the more similar the genes, the closer the relationship. By applying a known mutation rate for key genes and comparing the differences between species, it is possible to estimate approximately when lineages emerged and when individual species arose. (Redrawn from O'Brien & Johnson, 2007 'The evolution of cats' *Scientific American*.)

small species around the size of modern Wildcats to Lynxes. Just as *Pseudaelurus* was the progenitor of all sabretooths, *Styriofelis* led to all cat species living today (and many conical-toothed species which are now extinct). They evolved alongside the sabretooths, often with many members of both subfamilies occupying the same environment and presumably with the same complex inter-relationships we see among modern cat species today. Nine million years ago, the landscape that today surrounds modern Madrid had at least four species of now-extinct cats representing both subfamilies. Two species of Felinae, one Wildcat-sized (*Styriofelis vallesiensis*) and one Serval-sized (*Pristifelis attica*) must have occasionally fallen prey to two sabertooth species which were the size of a small Leopard (*Promegantereon ogygia*) and a Lion (*Machairodus aphanistus*). Fast forward to the early Pleistocene of East Africa and the Felinae branch had proliferated dramatically out of the shadows of their sabretoothed cousins. Around a million years ago,

the Cheetah, Leopard and Lion shared the African landscape with at least three species of large sabretooth cats. Sadly, we will never know how six large cats from the two great felid subfamilies interacted but their relationships must have been intriguing.

Approximately 11 million years ago, the genus *Styriofelis* began a rapid evolutionary radiation in Eurasia that would ultimately produce all lineages of living cats. By combining genetic analyses of living felids with the felid fossil record, the evolutionary relationships of modern cats and thus the composition of each lineage is now fairly well delineated. The actual number of cat species however is still surprisingly fluid. This book recognises 38 living felids, following the latest taxonomy adopted by the IUCN Red List (see p232) though taxonomy is not static and ever more sophisticated molecular analyses are likely to produce further revisions. Since 2006, genetic analyses have revealed the existence of 'cryptic

THE FELID LINEAGES

There are eight widely accepted lineages of modern cats which together make up the subfamily Felinae. As the most ancient and distinct lineage, Panthera is sometimes treated as its own subfamily Pantherinae but it is more correctly treated as a lineage with the Felinae.

Felis Lineage

The last lineage to diverge and therefore the youngest branch. Five closely related species in one genus, *Felis* with an African and Eurasian distribution. The taxonomy of the Wildcat and closely related Chinese Mountain Cat is unresolved. Some authorities classify the Chinese Mountain Cat as a Wildcat subspecies *Felis silvestris bieti* based on very limited genetic data which is disputed. There is some genetic and morphological evidence that the European Wildcat should be classified as a distinct species from African-Asiatic Wildcats. Finally, the domestic cat is occasionally elevated as a full species *Felis catus* although this does not have widespread support.

Prionailurus Lineage

Five species in two genera *Prionailurus* and *Otocolobus*, with a tropical and temperate Asian distribution. The evolutionary relationships of Pallas's Cat are poorly known, with evidence that it is intermediate to both *Prionailurus* and *Felis* but it is normally classified within the Leopard Cat lineage. There is some evidence for separating the Leopard Cat into two separate species where the dividing line is the narrow Kra Isthmus on the Malay Peninsula.

Puma Lineage

Three species in three genera which likely arose in North America and now have an Afro-Asian and American distribution. The Puma and Cheetah are not closely related to other large cats: they are essentially evolutionarily over-sized small cats that evolved larger bodies to occupy similar ecological niches to the *Panthera* genus.

Lynx Lineage

Four species in one genus with a temperate Eurasian and North American distribution. *Lynx* species are morphologically similar and distinct in all having bob-tails and ear tufts; it is unlikely that either feature has a strong selective advantage. This is among the most prey specialized of all felid lineages, particularly the Iberian Lynx and Canada Lynx.

Leopardus Lineage

Eights species in one genus with a Latin American (and barely USA) distribution. The eight members of this lineage diverged recently and are therefore closely related; hybridisation in the wild occurs between Southern Oncilla and Geoffroy's Cat and possibly between Northern Oncilla and Colocolo (see pp96-97). The Oncilla is now classified as two species and there is some evidence to separate out the isolated Central American population as a third species. The *Leopardus* lineage differs from all other cats in having 36 chromosomes rather than 38.

Caracal Lineage

Three medium-sized species in two genera with an African and Asian distribution. The Caracal is not closely related to the genus *Lynx* despite a similar appearance, suggesting perhaps a distant, common ancestor carried Lynx-like features or the incipient genetic variation for them.

Pardofelis Lineage

The second cat lineage to diverge following the Panthera branch and therefore one of the oldest. Three species in two genera with a Southeast Asian distribution. There is some evidence that the Marbled Cat should be reclassified as two separate species either side of the Kra Isthmus.

Panthera Lineage

The 'big cats'. The first lineage to diverge and therefore the oldest branch. Seven species in two genera, *Panthera* (the roaring cats) and *Neofelis* the Clouded Leopards. Most of the *Panthera* cats share the ability to roar, thought to be the result of unique modifications in the larynx and an elastic hyoid structure which supports it; and they cannot purr continually which is a feature of all other lineages. Snow Leopards and Clouded Leopards do not roar and both can purr.

species' of Oncillas (p96) and Clouded Leopards (p183). In both cases, populations across the range were traditionally regarded as a single species based chiefly on very similar morphology until genetic analysis revealed two very similar but genetically distinct species hiding in plain sight.

THE SOLITARY CAT

Early in felid evolution, the Felidae evidently settled on an essentially solitary lifestyle which, as far as the fossil record tells, has worked successfully for most species that have ever lived. It likely arose with the cat's ability to kill large prey without help. The feline template combines acute senses, hair-trigger reflexes, explosive muscular strength and a supple skeleton that promotes solo hunting; protractile claws and elastic wrists give tremendous control for grasping and handling large prey, while truncated, powerful jaws deliver a precise killing bite. Social carnivores like canids and hyaenas have more robust, less flexible bodies built for stamina to tire prey over long distances but which lack the cat's solitary killing prowess. A lone Puma is able to take down an adult Elk but it takes a few Wolves to do likewise.

Prey is also the main factor in the spacing patterns of cats. All cats attempt to secure sufficient resources for two fundamental requirements, survival and reproduction. Cats need water as well as suitable habitat for hunting, to avoid danger and to have cubs but these are usually secondary to their prey requirements. Female cats are *contractionists*. They occupy the minimum area required to provide for themselves and their cubs, which is dictated largely by their prey; its size, distribution in the landscape and how frequently it is replenished. Females can occupy small home ranges where prey is abundant, stable and uniformly distributed. Small ranges are more easily defended from competitors so they also promote increased territorial defence and little overlap between adjacent female ranges. Where prey is scarce, fluctuates dramatically or migrates, female ranges are larger, overlap is greater and territorial defence is reduced, often to a small,

exclusive core area (or areas).

Most felids show variation between these extremes according to the prevailing ecological conditions. A female Leopard living in very productive, prey-rich African woodland savannah can live her entire life in 10km² compared to a Kalahari Desert female that will have a home range 50 times larger. Equally, some species always tend towards one extreme. Snow Leopards only inhabit areas with dispersed prey at low densities so they always live in large home ranges at low densities. No female Snow Leopard lives in any habitat where she can meet her needs in a home range of 10km².

Male cats have ranges that are typically larger than predicted by food requirements alone. Males compete with other males for access to breeding females and they typically range over large areas in order to maximise the opportunities to encounter females; males are *expansionists*. They attempt to defend a territory from other males and monopolise females, so male ranges are usually larger than those of females. Where females maintain small, closely clustered ranges, a male can superimpose his territory over numerous female ranges with little overlap among males. Where females live at low densities in large home ranges, male ranges are typically larger and less exclusive with greater overlap between males.

The adults of most cat species undertake daily life alone (or are likely to, for poorly known species) but they are not asocial. Adults in the same area inhabit a complex social community of constant communication by scent-marks and vocalizations that allow familiar individuals to meet and rivals to avoid each other. Males and females come together to mate, and mothers are accompanied by the cubs of successive litters for most of their adult life but even among 'classically solitary' cats, adults often interact far more frequently and richly than is widely assumed. Unrelated adult Pumas in the Greater Yellowstone Ecosystem sometimes share large kills, perhaps because a large carcass can provide for many mouths and the costs of fighting over it are potentially too high. Similarly, male cats that apparently do little to provide for their cubs are still good fathers. Male Leopards and Tigers often spend time with their females and cubs, interacting amicably for extended periods including sharing kills.

Often dismissed as free agents that do little to raise cubs, male cats actually play a vital role. They patrol and defend the territory from immigrant males that would kill unrelated cubs if given the chance. Infanticide hastens the onset of oestrous in bereaved females, providing the new male with an accelerated window to produce his own offspring. Resident males repel such intrusions and provide mothers with the vital cloak of security that allows them to raise a litter to independence. In such a system, it is not so surprising that tolerance between familiar adults is manifested in regular social contact. No doubt, the same pattern – essentially solitary but

with tolerant, sometimes enduring social relationships – occurs in other cat species awaiting further study.

In a few felids, sociality is constant and complex. The Lion's extended families are built around a matriline of related adult females and their cubs that share a communal territory. Just as with solitary cats, Lioness range size is determined largely by prey requirements but rather than filling the needs of one mother and her litter, the range must meet the collective requirements of the entire female pride and its cubs. Male Lions live in coalitions that attempt to control as many female prides as possible which they defend from other coalitions – again, a scaled-up, highly social version of the basic felid socio-spatial pattern.

The Lion is the only cat that has developed the pride system. The reason lies, in part, again with prey. With both a rich diversity and high density of

So-called solitary cats are far more socially elastic than often portrayed. Here, two adult male Leopards consort with a female in oestrous, Sabi Sand Game Reserve, South Africa. The two males are almost certainly territorial neighbours and 'dear enemies'-rivals who know each other and who chose to be tolerant when the costs of fighting are high for both.

large herbivore species, African savannahs enable the formation of groups in large cats; simply put, there is enough to eat for big groups of big cats. Yet this does not automatically make pride-living the best strategy. In fact, the collective demands of feeding multiple mouths quickly overshadow the potential benefits of cooperative hunting. If food intake was the sole criteria, Lionesses would actually be better off alone or in pairs, particularly given that a Lioness is capable of killing all but the very largest prey species on her own. Indeed, it is that characteristic feline ability which probably fostered group living in Lionesses. In open savannah habitats, a large kill is a liability; it cannot be eaten quickly, is difficult to hide and is vulnerable to competitors. Lions evolved among all living large carnivores as well as three large sabretooth cats, and at least two extinct large hyaena species, any of which may have been able to dominate a single Lioness defending her kill. In such a highly competitive environment, it is better to share the kill with relatives who will help to defend it.

Ironically, group defence of kills would have created a further challenge for the ancestral Lioness. Female groups represent an extremely attractive resource to males and, just as with large kills, are likely to attract unwanted attention in open habitat. As incipient sociality emerged in Lions, so too did the risk of infanticide. Banding together would also have better enabled females to defend their cubs from foreign males. The evolution of the pride appears to be the Lion's response to acute competition over their cubs and kills. It did not arise in other felids presumably because the same combination of selective pressure and ecological opportunity did not exist. An individual Tiger, Leopard or Puma might gain a modest advantage from teaming up with conspecifics but that only makes evolutionary sense when that benefit outweighs the costs of group living. Presumably, none had the Lioness' persistent problem of defending very large, very obvious carcasses (or a similar ecological pressure), so remaining solitary is still the prevailing strategy for

the cat family. Aside from Lions, the only wild cat to form enduring social groups is the Cheetah, in which males may form coalitions for similar reasons to male Lions (though female Cheetahs are solitary; see pp 172–174).

THE SCIENCE OF STUDYING CATS

Wild felids are extremely challenging to study. Cats are generally rare, shy of people, and they often inhabit remote or inhospitable habitat. As a family, the Felidae does not lend itself easily to being observed, captured or monitored and many cat species are still very poorly known. The information in this book draws on thousands of scientific papers, reports and books written by researchers and naturalists spending months and sometimes years collecting data in the field. But how was that information collected? This section covers some of the main techniques we use to study wild cats.

Telemetry

Radio-telemetry has been a mainstay of wildlife research since the early 1970s. Until recently it relied largely on VHF (very high frequency) signals in which researchers use a receiver and directional antenna to detect a transmitter's signal and thus locate the animal. VHF radio-tracking relies on a clear line-of-sight between the transmitter (on the animal) and the receiver (with the researcher). Distance, dense habitat, mountainous terrain and even large electrical storms affect reception while species with large daily movements are often temporarily lost and large parts of their range might be entirely inaccessible, for example during the wet season. All impact the likelihood of finding a collared cat and therefore the quality of the data.

VHF radio-tracking is being replaced by collars with a GPS module, using the same technology as in car navigation systems. A GPS-collar automatically logs its own position as often as the researcher wants, provided the collar can communicate with the GPS satellite system. GPS collars store locations for later retrieval or they can be relayed remotely to the

researcher, via satellites or cellular phone networks (provided the collar is within range). GPS telemetry has major advantages over VHF telemetry in being able to automatically gather hundreds or thousands of accurate locations per collar and send them from the field to a laptop or cell phone anywhere in the world. Because GPS telemetry collects so much data, its results are more meaningful than similar efforts using VHF telemetry (see p181 for an example with Snow Leopards) and GPS collars are now sufficiently small to be useful for research of all cat species. Its main drawback is expense – a GPS collar costs 5–10 times as much as an equivalent VHF collar, and inexplicably high failure rates of the technology in the field. Every researcher who has deployed GPS collars on cats has experienced the terrific disappointment of collars failing to work as intended.

Telemetry is the mainstay of understanding felid spatial ecology. Locational data is used to calculate how much area a wild cat requires for its ecological needs – the home range (or territory, in case of those species which actively defend the range from conspecifics), as well as how it uses the area, for example, whether certain habitats or features in the landscape are preferred for certain activities such as raising cubs. If sufficient individuals (a large *sample size*) are telemetred, it also provides an insight into social and population ecology – how cats share the landscape with other members of their species, and how a population of cats behaves. Telemetry also provide an enormous amount of ancillary data on other aspects of felid behaviour including feeding and reproductive ecology, either by direct observation or gathering evidence after the event. Where direct observation is possible, think of radio-collared, vehicle-friendly Cheetahs in the Serengeti, researchers can view first-hand what cats hunt and kill, where they den their cubs and so on. GPS telemetry furnishes the same information even if the collared cat is never seen; location clusters direct researchers to possible kills or den sites that can be searched once the cat has moved off.

The greatest constraint of telemetry is the need to capture cats which is highly specialised, expensive and carries some risk to the animal. There is sometimes also concern about whether collars cause distress to cats, usually when collared animals are

A radio-collared Andean Cat in the Argentinean Andes, one of a handful of this species that has ever been collared. Provided that collars are small, lightweight and fitted carefully, cats completely ignore them.

viewed by tourists in parks though this is largely misplaced. Long-term monitoring shows that radio-collars do not influence survival, behaviour or reproduction, provided that collars are correctly fitted (which, most importantly, entails minimizing their weight). Similarly, the use of compact 'drop-off' devices automatically removes a collar without the need to re-capture the cat to recover it. Radio-collaring will continue to be an essential tool for wild cat research, especially for those species which have never been systematically studied. Nonetheless, it must be recognised that telemetry is not always the most suitable method for the question, and that advances in less invasive techniques provide extremely useful alternatives. These techniques are discussed next.

Camera-trapping

The use of camera-traps is now the most common research technique used by field researchers working on wild cats (and many other species). Camera-traps use a motion-sensitive sensor to trigger the camera, automatically taking a photo of whatever passes by. It is an enormously useful technique that overcomes many of the challenges of observing or capturing elusive species. Camera-trapping provides a wide variety of data types. A simple *inventory* helps to refine the distribution and status of cats; camera-trapping is the source of many recent new range records as well as, regrettably, losses in range. Repeated camera-trap surveys in areas of historic range without any evidence of the species probably indicates local extinction; the loss of Tigers from Cambodia, Lao PDR and Vietnam is a particularly unfortunate example.

Surveys that use many dozens or hundreds of camera-traps placed in a specific array permit scientists also to estimate the number of cats in a population. The process relies on recognising individual cats from photos; fortunately, spot, blotch or stripe patterns are unique to individuals, like a fingerprint. With enough photographs, *capture-recapture* statistical models use the relationship between the number of unique individuals photographed (or 'captured' by the camera) and how often each individual is photographed ('re-captures') to estimate the density of cats in an area. The technique has some constraints, for example, the survey must cover a sufficiently large area and continue for a long enough to sample a representative fraction of the population, but not all individuals in the area need to be photographed for an accurate result. Recently developed analytical models show promise for estimating population density even for cats which lack a unique pattern of spots or stripes such as Lions and Pumas. Most of the density estimates given in this book are calculated from data gathered during camera-trapping surveys and many of the photographs would not have been possible without camera-traps.

Repeated camera-trapping at the same site is very useful for monitoring changes in the status of cat populations either by detecting changes in the estimated density or by another analytical technique called *occupancy modeling*. Occupancy analysis uses camera-trap photos (or any evidence of a species including observations and tracks) to estimate the proportion of large survey areas in which the species is present and absent. Occupancy models employ powerful statistics to compensate for failing to detect the species when it actually is present but not observed during the survey, and they do not rely on being able to recognise individuals. Just as for changes in density, repeated camera-trap surveys at the same site can detect changes in occupancy; a decline in occupancy may signal a population under increased threat that warrants increased conservation effort.

Cats at the molecular level

Biologists working on cats have always collected their scats (field-worker shorthand for faeces), mainly to understand what they eat. Cat scat contains undigested signatures of their prey, hair, feathers, scales and claws, which, by comparing to a reference collection under a microscope, can often be identified to species. Much of the information on feeding ecology in this book comes from studying scats, especially of the lesser-known, poorly studied species where scats might be the closest a researcher ever comes to an encounter.

Until fairly recently, data on diet was as much as a researcher might hope to extract from old pooh but advances in molecular technology have opened extraordinary new opportunities for research. Every scat contains DNA of its owner in the naturally shed cells of the intestinal lining. Provided the scat sample is sufficiently fresh or well-preserved (for example, by sun-drying in very dry habitats), the process of isolating the DNA and identifying its origin is now routine. *Molecular scatology* can identify which cat species are present in the sampled area, how many individuals and their sex. Applying the same capture-recapture analyses as used with camera-trap data even allows researchers to estimate population size and density; each scat left by each individual substitutes seamlessly into the analysis instead of photos. Just as for DNA isolated from tissue, fur or blood, faecal DNA can used to analyse how populations are related and connected to each other, and their phylogenetic relationships; for example, a 2013 analysis of 601 scat samples from across the Puma's range shows three distinct groupings in North America, Central America and South America.

The sophistication and power of molecular analysis is now reaching the point that knowing which species and individual left a scat is just the beginning. Prey remains in the scat can be identified by the same process and the days of comparing hair samples under microscopes are beginning to wane. Researchers have even successfully isolated the DNA of internal parasites from scats. In the near future, molecular scatology will be able to furnish a complete genetic profile of the entire organism; the species, sex, individual, what it ate, whether it is carrying any parasites and even which viruses or bacteria it has recently encountered.

Using scats to learn about cats is particularly appealing because it does not require handling the animal (of course, samples of tissue and blood from captured or killed animals are routinely used for the same analyses). Hair holds some of the same advantages, at least in being able to identify species and individuals. Hair-traps – a sticky plate, barbed wire or wire brush – snag fur as the cat rubs against them although enticing felids to brush against them is a serious challenge. Ideally, DNA can be collected where cats chose to leave it; recently, researchers working in Sumatra demonstrated that it is possible to isolate Tiger DNA from urine-sprayed bushes used by the cats to demarcate territory.

Researcher David Mills sets a camera-trap during a study of the little-known African Golden Cat in Kibale National Park, Uganda. Effective camera-trapping relies on anticipating where cats will move; placing cameras indiscriminately in the landscape will produce few photos.

the wild cats

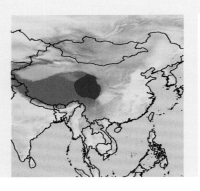

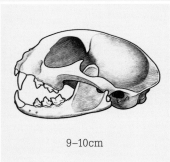

9–10cm

IUCN RED LIST (2008): Vulnerable
Head-body length 68.5–84cm
Tail 32–35cm
Weight 6.5–9kg

Chinese Mountain Cat

Felis bieti (Milne-Edwards, 1892)

Chinese Steppe Cat, Chinese Desert Cat

Taxonomy and phylogeny

The Chinese Mountain Cat is closely related to the Wildcat and a genetic analysis in 2007, based on a very small sample, reclassified the species as a Wildcat subspecies *Felis silvestris bieti*. However, the classification is disputed and requires further analysis. There are anecdotal reports that the species hybridises with domestic cats.

Description

The Chinese Mountain Cat resembles a large, long-limbed domestic cat. The fur is light yellow-grey in winter, darkening to tawny or grey-brown in summer with pale rufous or yellow-white underparts; it is locally called 'grass cat' due to its washed-out colour resembling dry grass. The species is not strongly marked except for a darkish, dorsal midline, and faint dark stripes and blotches are often apparent on the lower limbs, sides and nape, especially in the short summer coat. The tail is bushy and conspicuously banded with three to six dark rings and typically a dark tip; some individuals have a pale tip. The face has pale rufous streaks on the cheeks and forehead, and the ears are tipped with short tufts measuring 2–2.5cm.

Similar species The species most resembles the Wildcat, or a large domestic cat. Distant sightings through spotting scopes have emphasised the general impression of a robust sandy or ginger, somewhat Lynx-like appearance. Their range overlaps with the Pallas's Cat whose local names often mean 'mountain cat' (or similar) which can be misleading.

Distribution and habitat

Known only from Central China on the rugged, north-eastern edge of the Tibetan Plateau in Qinghai, Sichuan and Gansu Provinces. Records from elsewhere in China are equivocal and often based on skins in markets. Chinese Mountain Cats inhabit alpine grasslands, meadows, shrublands and forest edges from 2,500 to 5,000m. They may also occur in dense montane forest and desert but this is unconfirmed.

Feeding ecology

Chinese Mountain Cats are virtually unstudied in

Above: **A Chinese Mountain Cat listening for rodent prey in shallow snow. Hibernation does not occur in the cat family, and all felids continue to forage throughout winter.**

Left: **The Chinese Mountain Cat is the least photographed of all cat species in the wild. This adult was photographed in the remote Nian Bao Yu Ze mountains in China's Qinghai Province.**

the wild. A single study on diet showed that small rodents and lagomorphs, such as voles, mole-rats, hamsters and pikas, make up 90 per cent of the diet. Himalayan Marmots, Woolly Hares and Tolai Hares are common in their range, and are presumably also important prey. Birds, including pigeons, partridges and pheasants, are eaten. They apparently raid poultry especially during winter. Chinese Mountain Cats are reported to listen for mole-rats in their subterranean tunnels 3–5cm deep, and rapidly dig them out. The same technique would also work during winter when some small rodents are active in subnivean tunnels. They are reportedly nocturno-crepuscular.

Social and spatial behaviour

Chinese Mountain Cats are probably solitary but very little is known. Given that their preferred habitat is often quite open, refugia are probably a key factor in determining spatial patterns. They are known to rest and den in rocky terrain, under tree roots, in dense thickets and in the burrows of marmots and Eurasian Badgers. There is no information on home range size or density.

Like all cats, Chinese Mountain Cats have scent glands concentrated around the mouth and on the cheeks. Cheek-rubbing deposits an individual's signature smell, as an olfactory signal to both potential mates and rivals.

Reproduction and demography

A handful of records suggest seasonal breeding which is to be expected given the very harsh winters of their range. Male–female pairs are mostly observed in January to March, which is the likely breeding season, with two to four kittens born around May. One individual became independent at an estimated seven to eight months.

Mortality There are no records of natural deaths but putative predators include Grey Wolves, perhaps Golden Eagles and presumably domestic dogs.

Lifespan Unknown.

STATUS AND THREATS

Chinese Mountain Cats have a very restricted range and are thought to be naturally rare. They are killed for their fur which is mostly used by pastoralists; for example, to make traditional clothing articles such as hats or covers for small cushions. They are easily killed by leaving poisoned meat outside burrows and other rest sites. Although trade is largely domestic, hunting is **widespread** and pelts are common in fur markets and villages. Large-scale poisoning campaigns of rodents and lagomorphs are prevalent in pastoral regions of Central China in the belief that small rodents compete with livestock for grazing. Province-wide programmes ceased during the 1970s but poisoning is still widespread in the region at smaller scales, especially in Qinghai. The use of poison constitutes a grave threat by reducing prey populations and causing possible secondary poisioning of the cats. Poison is also used intentionally by fur hunters.

CITES Appendix II. Red List: Vulnerable. Population trend: Decreasing.

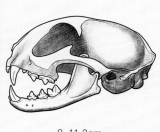

8–11.2cm

IUCN RED LIST (2015):
Least Concern (global)
European Wildcat: Vulnerable and
considered Critically Endangered
in Scotland

Head-body length ♀ 40.6–64cm,
♂ 44–75cm
Tail 21.5–37.5cm
Weight ♀ 2–5.8kg, ♂ 2–7.7kg

Wildcat

Felis silvestris (Schreber, 1777)

Asiatic Wildcat

European Wildcat

African Wildcat

Taxonomy and phylogeny

The Wildcat is part of the *Felis* lineage and is closely related to the Chinese Mountain Cat, which some authorities classify as a Wildcat subspecies, and more distantly to the Sand Cat with which it shared a common ancestor around 2.5 million years ago. There is considerable debate about the taxonomy and phylogeny of this species, with up to 19 described subspecies. The most robust and widely accepted of these based on morphological and genetic differences are the European Wildcat *F. s. silvestris*, African Wildcat *F. s. lybica* and Asiatic Wildcat *F. s. ornata* (sometimes referred to as Indian Desert Cat or Steppe Cat). Some subspecies have been treated as full species, and there is evidence that the European Wildcat warrants separation as a distinct species, *F. silvestris*. Under this classification, the Wildcats of Africa and Asia would be maintained as one species, *F. lybica*, with *F. l. ornata* as the subspecies inhabiting most of Asia, and a division of African-Middle Eastern populations into two subspecies: *F. l. cafra* in sub-Saharan Africa and *F. l. lybica* in the Middle East and presumably across North Africa (which lacked genetic samples in the study). The domestic cat arose from the Wildcat in the Fertile Crescent over 9,000 years ago. It is biologically the same species and occurs throughout the Wildcat's

The distinctive dark ear-tufts on this Wildcat help to identify it as the Asiatic form. Ear-tufts are largely unknown in Wildcats from other regions.

The familiar house cat

The African Wildcat *F. s. lybica* is the progenitor of the modern house cat, a process that began some 9,000–10,000 years ago in the Fertile Crescent with the rise of agriculture. Wildcats probably began associating with settlements as rodent populations boomed in fields and around grain stores; this 'self-domestication' was presumably fostered by people as the cats proved their value in controlling pests. Today's domestic cat is biologically still the same species as the Wildcat, and the two readily interbreed producing fertile kittens. The genetic distinctions between them are minor, though domestics carry a slightly different genetic signature, often considered sufficient to separate them as a subspecies, *F. s. catus*. Some authorities argue that domestic cats are unique by virtue of their long association with humans and treat them as a full species, *F. catus*. The domestic cat is an extraordinarily successful animal. Globally, at least 500 million cats are kept as pets and perhaps the same number live partially or entirely independent of people. As well as interbreeding with the Wildcat, the ecological impacts of the cat's success are staggering, primarily because the cat is such an efficient and adaptable predator. In the contiguous US, cats kill an estimated average of 2.4 billion birds and 12.3 billion small mammals a year; most mortality is by unowned cats (69 per cent of birds, and 89 per cent of mammals), but pet cats that roam outside account for the balance. On islands, where wildlife is often poorly adapted to predators, the impacts are even more extreme, and feral cats have contributed to 33 mammal, bird and reptile extinctions.

range where it is able to hybridise with wild individuals, further complicating the classification of this species (see Box 'The familiar house cat').

Description

The Wildcat's appearance is very similar to the domestic cat. Wildcats are generally slightly larger, longer-legged and more robust but feral or village cats in rural areas can be indistinguishable. Wild populations cluster into three major morphological groups (or morphotypes), reflecting the usual subspecies treatment, with considerable variation within each cluster and intergradations between them. The European Wildcat looks like a heavily built striped tabby with a pale to dark grey-brown coat, thickly furred tail and distinctive white chin and chest. The Asiatic Wildcat is pale isabelline to tawny-coloured and is the most heavily marked form, with dark brown to black dabs covering the body. Asiatic individuals often have small, dark ear-tufts that are rarely seen in other populations. The African Wildcat is sandy-grey to tawny-brown and generally lightly marked with indistinct spotting or striations on the body. Sub-Saharan individuals usually have richly coloured, brick-red ear-backs, distinguishing them from domestic cats. All forms have a striped tail ending in a dark tip, and have stripes on the upper legs; these form distinctive black 'armbands' on the forelegs in most sub-Saharan African populations. Piebald, ginger and black variants are the result of hybridisation with domestic cats.

Similar species Aside from the domestic cat, the

A European Wildcat in the snow. Wildcats occur above the snowline in much of their temperate Eurasian range but their distribution stops where deep snow is present for much of the year. Captive (C).

Chinese Mountain Cat is very similar. It lacks the obvious body spots typical of Asiatic Wildcats, though lightly marked Wildcat individuals are difficult to differentiate. Their ranges abut only at the north-western corner of the Tibetan Plateau in China. The larger Jungle Cat is sympatric across central Asia to India where the Wildcat is typically more heavily spotted and it lacks the distinctive white muzzle and short tail of the Jungle Cat.

Distribution and habitat

The Wildcat has an extensive distribution encompassing much of Eurasia and Africa. In western Europe, the range includes Scotland and continental Europe from Spain, France and Germany to Belarus, and east to the southern coast of the Black Sea and the Caucasus. It occurs across Africa but it is absent from West and Central African rainforest and from the open interior of the Sahara. It occurs patchily throughout the Middle East where it is mostly absent from the Arabian Peninsula interior, and it occurs in a broad swathe of south-western and central Asia from Iran along the east coast of the Caspian Sea to southern Kazakhstan, south-east to central India, and east to Central China and southern Mongolia. Its eastern distribution is bounded by the northern edge of the Tibetan Plateau and the Altai Mountains in Mongolia. It is absent from the interior of China's Taklamakan Desert.

The Wildcat occurs in an extremely wide variety of habitats. With the exception of dense, wet forest and open desert interiors, they occur in virtually all habitats with cover from sea level to 3,000m. They avoid very open habitat, exposed coasts, high, sparsely vegetated montane regions and areas with deep or continual snow. They readily occupy human-modified habitats of all kinds provided there is cover, including logged forest, farmlands, livestock pasture and plantations. They typically avoid intensively farmed habitat lacking vegetative cover.

An African Wildcat pursues rodent prey in the southern Kalahari, South Africa. Diurnal hunting is more likely in areas where Wildcats are not harassed or persecuted by people.

African Wildcats in northern Africa and the Middle East often have very pale sandy colouration with cinnamon-coloured markings, as this individual photographed in the Ouadi Rimé-Ouadi Achim Game Reserve, central Chad.

Feeding ecology

Wildcat diet is dominated by a combination of small rodents and, depending on relative availability, rabbits and hares. Most Wildcat populations have access to abundant rodents so the most common prey species include mice, rats, voles, hamsters, jirds, gerbils and jerboas. However, where lagomorphs are abundant or where rodent populations decline, hares and rabbits predominate (either generally or seasonally). In eastern Scotland, lagomorphs (mainly the European Rabbit) comprise up to 70 per cent of their prey. Young rabbits are the primary prey species during spring and summer, while adult rabbits are susceptible to the disease myxomatosis (which does not affect Wildcats) during winter, probably increasing their vulnerability to predation. Similarly in Spain, European Rabbits are the principal prey for Wildcats in low-altitude Mediterranean habitat, whereas they eat mainly small rodents in Mediterranean high mountain habitat where rabbits are absent. Very young or

small ungulates are sometimes recorded in Wildcat diet including ibex, chamois, Roe Deer, Red Deer, Wild Boar, duikers and gazelles, though most records are probably explained by scavenging. Small carnivores are sometimes killed, including records of Small-spotted Genets, Meerkats, Stone Martens, polecats and weasels, and cannibalism has been recorded rarely. After rodents and lagomorphs, birds make up the most important prey category for Wildcats, especially ground-foraging species such as doves, pigeons, partridges, sandgrouse, guineafowl, quails, sparrows and weavers. Wildcats in marshland and swamps are recorded taking a variety of waterbirds including coots, crakes and ducks, and there is a record from the Black Sea coast of a White-tailed Eagle (which might have been scavenged). Wildcats opportunistically eat a wide variety of reptiles, mainly small lizards and snakes, and there are records of large, venomous snakes (for example, vipers, puff adders and cobras), as well as amphibians, fish and invertebrates. Wildcats readily kill domestic poultry and, very rarely, neonate domestic goats and lambs typically no older than six to seven days. They drink daily when water is available but they often occur far from water in the Kalahari, Namib and Sahara, suggesting they can survive independently of standing water.

Wildcats hunt mostly on the ground, though they climb and swim very capably; they readily pursue prey in lower branches or tall brush and into shallow water or among inundated vegetation. They are chiefly nocturno-crepuscular but have flexible activity patterns depending on the region, season and the presence of competitors, predators or people; for example, Kalahari Wildcats are active for longer periods in the morning and earlier in the afternoon during the cold, dry season. Wildcats locate prey both by excellent sight and hearing, and stalk stealthily within striking range or react explosively when prey is flushed at close quarters, running it down or leaping to heights exceeding 2m. They also wait patiently at profitable sites, such as rodent burrows and especially for birds at waterholes or man-made watersources in arid habitat (for example, in the Kgalagadi Transfrontier Park, South Africa/Botswana, and Etosha National Park, Namibia).

Hunting success is known only from direct observation in the Kgalagadi Transfrontier Park where 3,676 hunting attempts were seen during a 46-month study, of which 2,553 (80 per cent) were successful. Female Wildcats had higher success (87 per cent) than males (69 per cent) because males hunted larger prey

A female African Wildcat and her kitten, southern Kalahari. The Kalahari's remoteness insulates its Wildcats from hybridisation with feral domestic cats.

more often with low success, and females hunted easily captured invertebrates more often than males; males and females were equally successful hunting the most important prey category, small rodents. In the same study, hunting Wildcats moved an average of 5.1km (range 1–17.4km) per night and captured an average of 13.7 prey items per night, with as many as 113 kills (many of invertebrates) in one night. A Wildcat in the Caucasus had 26 mice collectively weighing 0.5kg in its stomach. Wildcats readily scavenge including from dead livestock, and they sometimes cache larger kills by covering with sand, soil or leaf-litter.

Social and spatial behaviour

Wildcats are best studied in western Europe and southern Africa, and least known in Asia. The species is solitary and territorial, with characteristic feline territorial behaviours such as marking with urine and faeces, but the extent of territorial defence probably varies widely between different habitats. Range size varies considerably and generally follows the typical felid pattern of large male ranges overlapping multiple female ranges, but the sexes show little difference in some populations. Range

estimates include 1.7–2.75km^2 (females)–13.7km^2 (single male, Portugal); 1.75km^2 (average for both sexes, eastern Scotland with abundant lagomorphs); 3.5km^2 (females, average) to 7.7km^2 (males, average; southern Kalahari, South Africa); 8–10km^2 (both sexes, western Scotland with scarce prey); 11.7km^2 (average for both sexes, Saudi Arabia); and 51.2km^2 (single female, UAE). Density estimates from rigorous methods are surprisingly similar, probably coincidentally given that figures are available only for presumed medium to high-density populations of the species: 25 per 100km^2 (Kalahari, South Africa), 28 per 100km^2 (Mount Etna, Italy), 29 per 100km^2 (Jura Mountains, Switzerland) and 29 per 100km^2 (eastern Scotland).

Reproduction and demography

Wildcats breed seasonally in areas with extreme seasonality such as the Sahara and most of Europe, mating in winter to early spring and giving birth in spring to early summer. Elsewhere, kittens may be born year-round though birth peaks often coincide with prey flushes during or after the rainy season, for example in East and southern Africa. Females can have multiple litters per year; Kalahari females do

Two male European Wildcats in a territorial stand-off. Most encounters between neighbours are very ritualised so that rivals avoid escalation into serious aggression but fights occasionally result in serious injury.

not produce kittens during protracted shortages of rodent prey but can have up to four litters a year when rodent populations erupt. Gestation lasts 56–68 days. Litter size is typically two to four kittens, rarely up to eight. Weaning is at two to four months, and independence is 2–10 months. Both sexes are sexually mature at 9–12 months. Independent kittens have been observed visiting den sites and playing with their younger siblings and mother (Kalahari).

Mortality Rates are poorly documented across most of the range. Most mortality in studied populations is due to human factors; for example, 42–83 per cent of known deaths in two Scottish populations were from human causes, mainly shooting, snaring and roadkills. Known predators include large cats, large raptors, Red Foxes (of kittens), Honey Badgers (of kittens) and domestic dogs. Starvation of kittens and subadults contributes to low survival in severe northern winters.

Lifespan Up to 11 years in the wild and up to 19 in captivity.

STATUS AND THREATS

The Wildcat is widely distributed, common and tolerant of human presence. In some regions, for example much of savanna Africa, they have probably benefited from anthropogenic habitats, such as agriculture, which usually elevate rodent populations. However, potential benefits associated with people are accompanied by the most pervasive threat to Wildcats: hybridisation with domestic cats. Hybridisation is common among populations close to urban or rural human populations, for example in Scotland and Hungary where the Wildcat may soon cease to exist as a genetically pure form. Remote populations are relatively well insulated from hybridisation; for example,the Wildcat in the southern Kgalagadi Transfrontier Park (South Africa and Botswana) is still genetically pure. Hybridisation is accompanied by other anthropogenic threats that have led to significant population declines or range loss. The presence of domestic cats is particularly critical for populations already heavily impacted by more conventional threats, for example in much of western Europe where persecution as part of predator-control campaigns and habitat loss was formerly very widespread. Similarly, extreme habitat conversion across much of the Wildcat's Indian range has produced population declines in which hybridisation may act as the final blow. The Asiatic Wildcat was formerly trapped in large numbers for its spotted fur across much of Asia. There is relatively little international trade now, though it is still killed for fur in China, which also conducts rodent and pika poisoning campaigns in parts of the species' Chinese range, the effects of which are unknown. CITES Appendix II. Red List: Least Concern (European Wildcat: Vulnerable and considered Critically Endangered in Scotland). Population trend: Decreasing.

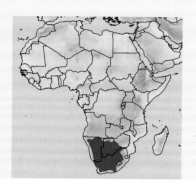

7.2–8.7cm

IUCN RED LIST (2008): Vulnerable

Head-body length ♀ 35.3–41.5, ♂ 36.7–52cm
Tail 12–20cm
Weight ♀ 1.0–1.6kg, ♂ 1.5–2.45kg

Black-footed Cat

Felis nigripes (Burchell, 1824)

Small-spotted Cat

Pale form

Taxonomy and phylogeny

The Black-footed Cat is classified in the *Felis* lineage where its closest relative is thought to be the Sand Cat, though the data are poor. Two subspecies are usually recognised: *F. n. thomasi*, restricted to the Eastern Cape region, South Africa, and *F. n. nigripes*, which occurs everywhere else in the range. The distinction is based on differences in pelage and size that vary along a cline (see Description) and have not been validated by genetic analyses.

Dark form

Description

The Black-footed Cat is one of the two smallest cats, slightly larger on average than the Rusty-spotted Cat, with a stockier thickset body and relatively short legs. Male Black-footed Cats are heavier than male Rusty-spotted Cats. The body fur is light tawny to cinnamon-buff and varies from pale to dark along a north–south cline. Southern individuals (*F. n. thomasi*) are richly coloured with bold black spots and blotches whereas northern individuals (*F. n. nigripes*) are paler with coffee-brown to rust-red-tinged markings. The differences are obvious only at the extremes of the range and they disappear around Kimberley, South Africa, where the population has characteristics of both forms. Individuals throughout the range have large dark blotches forming a yoke across the upper chest and bands around the upper limbs. The backs of the forefeet and hindfeet are black.

Similar species The Wildcat is sympatric in the Black-footed Cat's entire range and also has dark fur on the back of the hindfeet, leading to possible confusion in fleeting sightings, but the Wildcat is larger and is plainly coloured without distinct body markings. The Serval is very similarly coloured and marked to the Black-footed Cat but is much larger with a tall, long-legged build and very large ears.

Distribution and habitat

The Black-footed Cat is endemic to southern Africa, where it occurs primarily in South Africa, Namibia and Botswana (the last lacks recent records for much of the putative range). It is recorded from extreme north-west and southern Zimbabwe close to the South African border, and similarly is thought to occur marginally in extreme south-eastern Angola given there are records on the Namibian border. It has not been recorded from Lesotho, Mozambique or Swaziland. The Black-footed Cat is a specialist of open, dry habitats with cover, especially short-grass

Right: **Some raptors follow small carnivores as they hunt, to pick off flushed prey before the predator can pounce. Such a hunting association has been observed between the Marsh Owl and Black-footed Cat in South Africa; it is thought only the owl benefits.**

Below: **An adult Black-footed Cat on the hunt in typical high grass habitat, Benfontein Nature Reserve, South Africa.**

savanna, Karoo scrub, open savanna woodlands with sparse tree cover, and vegetated semi-desert. Termitaria or burrows made by other species appear to be crucial for shelter; 98 per cent of documented rest sites during a study in central South Africa were inside abandoned Springhare burrows. Black-footed Cats may be absent from habitats lacking these or similar refugia. They do not occur in open, hyperarid areas of the Kalahari and Namib deserts dominated by sand or gravel plains, which lack suitable vegetative cover and refuges. Black-footed Cats occur in some anthropogenic habitats, such as livestock areas, but they are intolerant of intensive modification including most agriculture and plantations. Records of the species occur from sea level to around 2,000m.

Feeding ecology

The Black-footed Cat is a very active and opportunistic hunter, primarily of very small mammals and small ground-foraging birds. In the only comprehensive study of the species, in Benfontein Nature Reserve, central South Africa, various species of shrews, mice and gerbils comprised 73 per cent of prey items (of 1,725 records) and 54 per cent of all biomass consumed. Birds were the next most important category, making up 16 per cent of prey items and 26 per cent of biomass eaten. Twenty-one bird species and their eggs were taken, mainly various larks, cisticolas, coursers and buttonquail, and the largest species killed was the Northern Black Bustard (~0.7kg). Birds are often taken in flight, in leaps up to 1.4m high. Black-footed

Black-footed Cats occupy habitats with relatively little terrain and structure. Burrows in termite mounds, usually created by Aardvarks or Spring Hares, provide essential refugia from predators and the elements.

Right: **A grooming Black-footed Cat, showing the characteristic black fur that covers the underside of the foot from the ankle to the toes. The feature is distinct in brief night-time sightings as the cat flees for cover.**

Below: **The blue eyeshine in this Black-footed Cat comes from the *tapetum lucidum*, a layer of reflective cells that bounces light back to the retina to augment vision in low light.**

Cats can overpower prey very close to their own weight, for example Cape Hare (1.5kg) and young Red Rock Rabbit (1.6kg), which are the largest prey species recorded. Males have been observed unsuccessfully attacking resting newborn Springbok lambs (~3kg) which are able to thwart attacks simply by standing up. Suricates, Yellow Mongooses and Cape Ground Squirrels are occasionally killed. Reptiles, amphibians and invertebrates (especially termites, locusts and grasshoppers) are regularly eaten, though they contribute minimally to intake, for example 2 per cent of biomass in Benfontein Nature Reserve. While Black-footed Cats are easily able to take poultry, there are no records, perhaps because they appear to avoid human settlements. Historical accounts of them killing goats and sheep are not

credible. Black-footed Cats are independent of drinking water but they drink where water is available.

The Black-footed Cat forages almost exclusively on the ground and nocturno-crepuscularly. They are extremely active hunters and spend around 70 per cent of the night moving in search of prey in all weather conditions, and at temperatures of -10°C–35°C. They use three distinct hunting techniques. When 'fast-hunting', they rapidly and randomly bound through cover to opportunistically flush prey (especially birds) from hiding. During 'slow-hunting' they weave silently and deliberately around cover, using their excellent hearing to locate prey. Finally, they also wait silently in ambush at rodent burrows for up to two hours. Most prey species, including lagomorphs, are killed by a head or neck bite. Snakes are repeatedly struck on the head until sufficiently stunned or tired for the cat to deliver a neck bite. Sixty per cent of hunts succeed (Benfontein Nature Reserve) with 10–14 rodents or birds caught per night, averaging a kill every 50 minutes. They consume around 20 per cent of their body weight each night. Surplus food is cached in shallow Aardvark diggings or carried to hollow termitaria. They also scavenge, for example from Springbok carcasses.

Social and spatial behaviour

The Black-footed Cat is solitary and territorial. Males have larger ranges than females and overlap the ranges of one to four females. Range overlap within the sexes is generally minimal, especially between neighbouring males. Female ranges overlap more, with up to 40 per cent comprising shared areas, possibly because related females are likely to be neighbours. Territories are demarcated mainly by urine-marking. This escalates dramatically during the mating season when males spray up to 585 times a night, accompanied by loud vocalising as they search for females. Males guard oestrous females and fight other males. Ranges average 8.6km^2 (up to 12.1km^2) for females, and 16.1km^2 (up to 20.2 km^2) for males. Density is not well known but 16.7 cats per 100km^2 is estimated for high-quality habitat in central South Africa (Benfontein Nature Reserve).

Reproduction and demography

Breeding occurs year-round in captivity but is seasonal in the wild. Births coincide with rain and prey flushes during the southern African spring and summer (September to March) and do not occur during the cold, prey-poor winter. Wild females may have up to two litters per year usually when the first litter dies; a captive female had four litters in one year. Oestrus lasts only 36 hours, probably an adaptation to reduce vulnerability to predators while consorting and mating in the open. Gestation is 63–68 days. There are one to four kittens, usually two in the wild; a litter of six reported in captivity was never verified. Kittens are kept in hollow termitaria or Spring Hare burrows. They are weaned by two months and independent at three to four months but typically remain in the natal range for much longer. Sexual maturity in captivity occurs at the earliest at seven months (females) and nine months (males).

Mortality Poorly known. Predators of adults and kittens include Black-backed Jackals, Caracals, domestic dogs and large eagle owls. There is a record of an adult male dying due to a burrow collapse, and a kitten died when caught in the open during a severe hailstorm.

Lifespan Up to eight years in the wild, up to 16 in captivity.

A female Black-footed Cat waiting patiently outside a rodent burrow, adopting a sit-and-wait hunting strategy that expends little energy.

STATUS AND THREATS

The Black-footed Cat is very inconspicuous and its status is difficult to assess. Based on indirect measures such as roadkills and the low frequency with which they are killed during predator-control campaigns and by fur trappers, they are probably naturally uncommon to rare. Large parts of their range, for example most of the Kalahari, lack verified or recent records; the current estimate of distribution may include large areas where the species does not occur. The main threat to the Black-footed Cat is the expansion of agriculture and livestock grazing into semi-arid areas. Overgrazing by livestock and the use of agro-pesticides impacts the populations of rodent prey and of insects (which impacts bird populations) and may secondarily poison Black-footed Cats. Indiscriminate use of poisons, traps and dogs to control jackals and Caracals in South Africa and Namibia is also a significant threat.

CITES Appendix I. Red List: Vulnerable. Population trend: Decreasing.

8–9.5cm

● **IUCN RED LIST (2008):**
Near Threatened

Head-body length ♀ 39–52 cm,
♂ 42–57 cm
Tail 23.2–31cm
Weight ♀ 1.35–3.1 kg, ♂ 2.0–3.4 kg

Sand Cat

Felis margarita (Loche, 1858)

Taxonomy and phylogeny

The Sand Cat is classified in the *Felis* lineage where
it is closely related to the Wildcat and Black-footed
Cat. It has sometimes been classified with Pallas's
Cat in the genus *Otocolobus* based on morphological
specialisations (mainly in the auditory region of the
skull) but genetic analysis places it clearly in *Felis*.
Four subspecies are usually described, based broadly
on the species' putative discontinuous distribution,
but further molecular analysis is required to assess
their validity.

Description

The Sand Cat is a very small, strikingly pale cat with
a flat, broad head topped by very large ears. The fur
is pale sand coloured, light grey or rich golden-sandy,
often finely speckled with a faint saddle of
interspersed black and silvery hairs over the nape,
shoulders and flanks. The body is marked with
partial, longitudinal dark bands and spots that are
typically indistinct or absent entirely, but are dark
and obvious in a small number of individuals. The
markings become more distinct on the limbs, as

prominent black 'armbands' on the upper forelegs, and slightly less obvious striations on the hindlegs. The tail is faintly banded, resolving to distinct dark stripes towards the tip, which is black. The winter coat is very dense and long in central Asia (and probably elsewhere in the range given the extremes in temperature). Sand Cat feet are densely covered in fine, dark fur, probably for traction and insulation on loose, hot sand.

Similar species The closely related Wildcat is especially pale coloured and faintly marked in parts of the Sand Cat's range, for example in overlapping areas of Algeria, Niger and the Arabian Peninsula where the two species are easily confused. The Sand Cat's enormous ears and short legs are characteristic and it has a distinctive rolling gait when moving quickly compared to the bounding action of the Wildcat.

Distribution and habitat

The Sand Cat has a discontinuous distribution, with apparently separate populations in central Asia from southern Kazakhstan to central Iran; the Middle East and the Arabian Peninsula; and in North Africa west of Libya (from which there are no records). It was recorded for the first time in Chad, in the Ouadi Rimé-Ouadi Achim Game Reserve, in 2014; this suggests the range is more continuous across North Africa. The Sand Cat was formerly recorded in four countries where the most recent records are at least 10 years old and its current presence is uncertain: Egypt, Israel, Yemen and Pakistan. It is unclear if the putative gaps in the range represent true absence or, more likely, a lack of records; the species is extremely elusive to visual surveys in most of its range. It was definitively recorded from Syria and Iraq only in 2001 and 2012 respectively.

The Sand Cat is a desert specialist, capable of inhabiting true desert with rainfall as low as 20mm per year. Sand Cats inhabit a variety of sparsely vegetated sandy or stony desert (*hammada*) habitats, and arid shrub-covered steppes. Within these

A captive Sand Cat in its dense, winter coat. This individual has unusually prominent striping on the body, seen relatively rarely in the species. (C).

Right: **The thickly furred feet of Sand Cats act like the desert equivalent of snowshoes, providing traction and insulation on hot, loose sand (C).**

Opposite: **Except for birds, virtually all Sand Cat prey uses burrows and other sub-surface refuges. Sand Cats are well adapted to rapidly excavate subterranean prey such as scorpions, spiders, reptiles and rodents.**

Below: **Sand Cats are known to cache killed prey with a sandy covering. Caching may provide a larder in case of food shortages in unpredictable desert habitat, although details of the behaviour are poorly understood.**

habitats, they are usually absent from areas dominated by shifting sand dunes that cannot sustain any vegetation and, at the other extreme, from heavily vegetated areas such as in deep valleys and thickets (for example, where Black Saxual scrub forms stands of low 'forest' in central Asia). Sand Cats occur in areas with extraordinary temperature ranges, for example from 45°C in summer to -25°C in winter in the Karakum Desert, Turkmenistan.

Feeding ecology

The Sand Cat subsists on desert-adapted small vertebrates. The low diversity of its habitat means the number of species recorded in the diet is relatively few compared to most felids, but this should not be misinterpreted; it is an opportunistic and highly successful, generalist hunter. The diet is generally dominated by small mammals, particularly spiny mice, gerbils, jirds, jerboas and ground squirrels, as well as hamsters and young Tolai Hares and Cape Hares. Birds and reptiles are also important prey, potentially outranking mammals as the most important category depending on availability; for example, during seasonal fluctuations in rodents. Sand Cats are recorded taking sparrows, larks, jays, doves, partridges and sandgrouse; woodpeckers are rarely recorded from the central Asian range. A wide variety of lizards and snakes are eaten including highly venomous horned and sand vipers (genus *Cerastes*). Invertebrates, including scorpions and spiders, are readily eaten. They sometimes enter settlements to kill poultry and are reported to drink camel's milk left sitting in gourds. Sand Cats are

independent of drinking water but they drink where water is available, and are often found near water sources presumably because these attract prey.

Sand Cats forage at night during the hottest periods of the year, but incidental observations across the range suggest diurnal activity is common at other times, especially during winter in areas where temperatures at night drop below freezing. Hunting behaviour is not well known; they deal with vipers by landing a succession of rapid blows to the head until it is safe to dispatch the snake with a bite to the neck or skull. Sand Cats are capable of very rapid digging to excavate burrowing prey and they sometimes cache food with a covering of sand. It is unknown if they scavenge.

Social and spatial behaviour

The Sand Cat is poorly known with very limited data from telemetry only from Israel and Saudi Arabia. They are solitary and presumably maintain enduring ranges, though it is unknown if they defend territories from other adults. They move long distances while foraging, with up to 10km recorded in a single night. The low productivity of their habitat and wide-ranging movements would suggest that ranges are large and densities are low. During the

day, Sand Cats rest in burrows or among the wind-exposed roots of dense brush, which they further excavate. Based on a small number of observations, the ranges of males in southern Israel overlapped one another and one male's range was 16km² (probably an underestimate). The only available density estimate comes from the same study where 11 cats were captured in 375km² – that is, three cats per 100km² – but this is an underestimate given other cats were likely living in the area and not captured.

Reproduction and demography

Sand Cats in captivity breed year-round, though limited information from the wild suggests that breeding is seasonal, at least in some parts of the range, which is plausible given the environmental extremes of Sand Cat habitat. In the Sahara and central Asia, mating apparently occurs November–February and kittens are born January–April. A litter is recorded from Pakistan in October, which has given rise to the suggestion that two litters a year are produced, though this more likely indicates aseasonal breeding in some parts of the range. Gestation in captivity lasts 59–67 days. The litter size is typically two to four kittens, but they are capable of producing large litters, and up to eight is recorded from captivity. Weaning occurs around five weeks, and independence occurs from four months. Sand Cats are sexually mature at 9–14 months (captivity).

Mortality There are few records of natural mortality; the species often occurs in habitats with few natural enemies. There is one record of two

Golden Jackals killing a Sand Cat in Niger. Large raptors, including Pharaoh Eagle Owl, Eurasian Eagle Owl and Golden Eagle, are possible predators, especially of kittens. Population crashes are recorded in central Asia during prolonged snowy winters, mainly from starvation due to heavily reduced rodent numbers.

Lifespan Unknown from the wild and up to 14 years in captivity.

Below: **Record litters of eight kittens from captivity suggest the Sand Cat is able to increase reproductive output when environmental conditions for breeding are optimal. (C).**

STATUS AND THREATS

Sand Cat status and the effects of threats are poorly known. The species appears to occur at low densities everywhere it is found and is thought to be naturally rare. Arid ecosystems are particularly susceptible to degradation and human presence, especially in the form of expanding cultivation and livestock grazing. The associated presence of feral domestic cats and dogs may be a threat with the potential of competition, disease transmission and predation (by dogs), though effects on Sand Cats are unknown. Sand Cats are sometimes caught near oases and human settlements in traps intended for jackals and foxes, and they are persecuted for real or suspected poultry killing. Localised threats associated with the presence of people are perhaps balanced to an extent by great swathes of Sand Cat habitat being very remote and largely uninhabited, insulating them from human activities. Desertification and prolonged droughts affect much of the range, which potentially increases Sand Cat habitat and depopulates areas of humans, but both factors often result in habitats dominated by shifting sand, with little vegetation or prey, which Sand Cats avoid.

CITES Appendix II. Red List: Near Threatened. Population trend: Unknown.

9.8–14cm

● **IUCN RED LIST (2008):**
Least Concern

Head-body length ♀ 56–85cm,
♂ 65–94cm
Tail 20–31cm
Weight ♀ 2.6–9.0kg, ♂ 5.0–12.2kg

Dark form

Pale form

Jungle Cat

Felis chaus (Schreber, 1777)

Swamp Cat, Reed Cat

Taxonomy and phylogeny

The Jungle Cat was once thought to be closely related to the lynxes due to superficial physical similarities, but there is no dispute that it belongs in the *Felis* lineage. It is thought to have diverged early in the lineage over three million years ago and is perhaps most closely related to the Black-footed Cat, though the genetic data are poor. Up to six subspecies have been described based largely on superficial differences, especially in pelage, which varies within and between populations. Most subspecies are unlikely to be valid and a modern molecular analysis of subspeciation is overdue.

Domestic cats in rural areas, for example, in India and Indochina, often strongly resemble Jungle Cats, raising the possibility of widespread hybridisation (which is known from captivity). There is at least one record from India of a wild male Jungle Cat fathering kittens to a female domestic cat.

Description

The Jungle Cat is the largest of the *Felis* cats, with a tall, slender build, long legs and a fairly short, banded tail measuring around a third of the body length. Jungle Cats in the west and north of the range are believed to be the largest – based on a few specimens, Jungle Cats in Israel are 43 per cent heavier than those from India. The Jungle Cat's head is relatively compact with large, triangular ears topped with a short, sometimes indistinct dark tuft; the tuft is usually obvious in kittens. Jungle Cats are uniformly coloured, typically light to dark tawny with greyish, gold or rusty tones. The body is faintly marked with indistinct stippling or spots that are entirely absent in some individuals. The lower limbs and tail are more distinctly marked with dark blotches and banding. Young kittens are often more strongly marked, with distinct blackish dabs covering the body at birth that fade rapidly. The Jungle Cat's face is lightly marked with indistinct cheek and forehead stripes and, typically, a distinctive, bright white muzzle and chin. Individuals in temperate regions are often more richly coloured and more heavily marked; summer coats are highly variable with rich orange to coffee-brown tones while silver-grey winter coats are recorded in the extreme north of the range. Melanistic individuals are reported from India and Pakistan.

Similar species The Jungle Cat's tall, leggy build and short tail resembles the Caracal, which is sympatric from India to Turkey; Caracals have very distinct long ear tufts and lack the leg and tail markings of Jungle Cats. Young Jungle Cats are easily mistaken for Wildcats or feral domestic cats. The Jungle Cat's white muzzle is usually distinctive.

Distribution and habitat
The Jungle Cat has an extensive but patchy range from Vietnam and southern China to western Turkey and Egypt. The species' stronghold is South Asia where it occurs throughout India, Bangladesh, much of Sri Lanka and through the Himalayan foothills from Pakistan to Burma. In Indochina it occurs from southern China to southern Thailand and Cambodia, with large gaps in between where it has been extirpated, is very rare or simply unknown. It occurs patchily in south-western and central Asia from Turkey to southern Kazakhstan, western Afghanistan and southern Iran. In Africa, the Jungle Cat is restricted to Egypt on the Mediterranean coast, along the Nile River Valley from the delta to

A large male Jungle Cat showing the species' characteristic leggy build and short, banded tail. Males are taller at the shoulder and more heavily built than females.

Right: **Jungle cats are very at home in water. They swim powerfully and hunt in the shallows for aquatic prey including a wide variety of fish.**

Aswan, and in a scattering of oases west of the Nile.

Despite its name, the Jungle Cat avoids dense, forested habitats and is probably entirely absent from closed canopy rainforest. The species' alternative names, Swamp Cat and Reed Cat, are more appropriate as it strongly prefers well-watered and dense reedbeds, grassland and scrubland associated with swamps, wetlands, marshes and coasts. They occur in oases and river valleys in arid habitats, for example in Egypt and southern Iran, and they inhabit cleared areas and grassland or scrub patches in moist forest. In Indochina, the species mainly inhabits deciduous woodland and open forest with rivers, floodplains and other scattered water sources. The Jungle Cat tolerates converted landscapes with water, cover and prey. They can be found in close association with people in agricultural habitats that foster dense rodent populations, for example in sugarcane fields, rice paddies and irrigated pastures. They also occur around aquaculture ponds, and in various open forest plantations, especially those with irrigation canals. The Jungle Cat occurs from sea level to 2,400m in the Himalayan foothills.

Below: **Although often confused with feral domestic cats, this young Jungle Cat already shows the species' characteristic enlarged ears and pale lower face.**

Feeding ecology

The Jungle Cat feeds primarily on small mammals weighing less than 1kg, mainly mice, rats, gerbils, jirds, jerboas, voles, ground squirrels, moles and shrews, as well as Muskrats (weighing up to 2kg; introduced in parts of the range) and hares. A single Jungle Cat is estimated to eat 1,095–1,825 small rodents per year in semi-arid western India (Sariska Tiger Reserve). Jungle Cats are relatively powerfully built and occasionally take larger mammals including Coypu (weighing 5–9kg), though a large adult Coypu was successfully observed to repel an attack in Russia (the age and size of the Jungle Cat was not reported). Jungle Cats occasionally take the neonates of gazelles and Chital, and an adult cat killed captive subadult Mountain Gazelles held in large pens. Wild Boar piglets are reported as prey from the western shore of the Caspian Sea, though these were possibly scavenged; only unaccompanied, very young juveniles would be vulnerable. Birds are the second most important prey category including small species (especially grassland sparrows and larks), waterfowl, francolins, pheasants, peafowl, jungle fowl and bustards. Jungle Cats in Uzbekistan are recorded taking Western Marsh Harriers. They also eat reptiles, amphibians and a wide variety of fish. Tajikistan cats are recorded to have eaten large quantities of Russian Silverberry fruits during a lean winter. Jungle Cats prey on domestic poultry and are sometimes blamed for killing small stock; this is possible, though the records are equivocal.

The Jungle Cat forages mainly on the ground in

dense cover, often along the water's edge or among inundated vegetation. It is an excellent swimmer that crosses long stretches of open water to small islands and reedbeds, and it actively takes fish and waterfowl in the water. They are reported to excavate Muskrats from their small 'push-ups' (similar to beaver lodges). Foraging is thought to be cathemeral with a high degree of crepuscular and diurnal activity in sites that are well protected from people. The Jungle Cat readily scavenges, including removing animals from human traps and eating from the kills of other carnivores such as Golden Jackal, Grey Wolf, Asiatic Lion and Tiger.

Social and spatial behaviour

The Jungle Cat is very poorly known. There are no studies based on telemetry or long-term observation but limited information indicates it follows a characteristic felid socio-spatial system. Adults are solitary, and typical feline scent-marking and vocalisation suggest the maintenance of exclusive core areas. Based on incidental observations in Israel, male home ranges overlap several smaller ranges of females. There are no rigorous density estimates but they are locally common in suitable habitat in South Asia.

Reproduction and demography

This is poorly known from the wild. The species is thought to breed seasonally, which is credible for parts of the range with climatic extremes, for example Egypt, the Middle East and the extreme north of the range (Kazakhstan through southern Russia) from where most observations exist. Records of mating from these areas occur mostly in November–February and kittens are born December–June. Gestation in captivity is 63–66 days. Litters are usually two to three kittens, exceptionally up to six. Kittens are independent at eight to nine months old. Sexual maturity in captives occurs at 11 months (females) and 12–18 months (males).

Mortality There are few records of natural mortality but they are presumably killed occasionally by larger carnivores; Golden Jackals occur at high densities in much of their range and may be a threat to kittens. A dead Jungle Cat in Pakistan was found entwined with a large dead Indian Cobra, with evidence of a prolonged struggle. Humans and dogs frequently kill kill them in anthropogenic landscapes.

Lifespan Unknown in the wild and up to 20 years in captivity.

A Female Jungle Cat and her three kittens, Khijadiya Bird Sanctuary, Gujarat, India. Khijadiya is one of the few sites where it is possible to view wild Jungle Cats for extended periods.

STATUS AND THREATS

The Jungle Cat is widespread and is able to tolerate agricultural and settled landscapes where it is probably the most common felid in some regions, particularly in rural areas of South Asia. It is uncommon to very rare in Indochina, south-western Asia and central Asia. It was formerly common in its Egyptian range, though its status there is now uncertain. Despite its ability to live close to people, the species is threatened by the rapid conversion of wetland habitats for human settlement and agriculture. In Indochina and much of South Asia, this is compounded by very widespread, unselective snaring to which the Jungle Cat is particularly vulnerable given its preference for open habitats that are very accessible to people. In the northern parts of the range, it is trapped for fur and it is hunted heavily around Coypu and Muskrat fur-farms in the former USSR republics. They are persecuted for poultry-raiding and have declined in agricultural areas where unselective trapping and poisoning are prevalent. CITES Appendix II. Red List: Least Concern. Population trend: Decreasing.

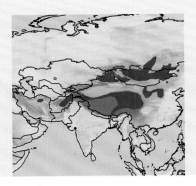

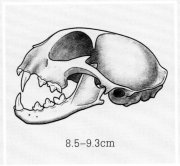

8.5–9.3cm

IUCN RED LIST (2015):
Near Threatened

Head-body length ♀ 46–53 cm,
♂ 54–57
Tail 23–29cm
Weight ♀ 2.5–5kg, ♂ 3.3– 5.3kg

Pallas's Cat

Otocolobus manul (Pallas, 1776)

Manul, Steppe Cat

Taxonomy and phylogeny

Pallas's Cat is classified as the only species in the genus *Otocolobus* on the basis of its unique morphology and genetic data that show it is related to both the *Felis* and *Prionailurus* lineages. There is some evidence that its closest (but still rather distant) relative is the Leopard Cat, and so Pallas's Cat is often grouped in the *Prionailurus* lineage but the issue remains unresolved. Three subspecies are usually described but further molecular analysis is required to assess their validity.

Description

Pallas's Cat is a very distinctive stocky, heavily furred small felid about the size of a squat domestic cat. The broad head is very distinctive with a flat forehead and a wide face, enhanced by dense cheek sideburns and large ears set on the side of the head. The face has characteristic paired dark stripes usually with white fur in between running from the eyes to the cheeks, and the forehead is distinctly marked with small black spots. The dense fur is silvery-grey to rufous-grey, and largely unmarked on the body or sometimes with faint vertical striping. The winter coat can be very long and dense with a pale, frosted appearance; the summer coat tends to be darker with richer, often reddish, tones and more obvious markings on the body. The bushy tail is banded with narrow stripes ending in a dark tip. Pallas's Cat's colouration provides excellent camouflage in open, rocky habitat. It is poorly adapted for running and on detecting a possible threat such as a distant predator it freezes and flattens itself to the ground, which is an effective method of concealment. Pallas's Cat weights fluctuate considerably depending on the season. In Mongolia, radio-collared females lost an average of 23 per cent body mass after raising kittens in late summer (July to September) and, after recovering weight, had a second low at the end of winter (February to March). Male weights peaked just before the winter mating season in November and lost an average of 22 per cent mass by the end of the mating season in March.

Similar species Pallas's Cat is a very distinctive felid and unlikely to be confused with other species except perhaps the Chinese Mountain Cat, which is sympatric only in Central China.

Distribution and habitat

Pallas's Cat has a wide but patchy distribution across Eurasia's cold steppes. Its main stronghold is in Mongolia, neighbouring areas in Russia and across much of China with the distribution becoming progressively discontinuous further west. It is thought to occur in relatively large contiguous areas in eastern Kyrgyzstan, eastern Kazakhstan and

Pallas's Cats' fur is typically a shade of light to dark grey but red variants occur, especially in central Asia. This captive individual is a particularly richly coloured example (C).

northern Afghanistan, and in more isolated fragments in western Pakistan, northern India and northern Iran. Its range is very limited, fragmented or uncertain in Azerbaijan and Turkmenistan. The species was camera-trapped for the first time in both Bhutan (in Wangchuck Centennial Park) and Nepal (in the Annapurna Conservation Area) in 2012, approximately 100km south and 500km west respectively of the southernmost record on the Tibetan Plateau in China. It is likely the species occurs continuously along the Himalayan range.

Pallas's Cats live in cold, arid habitats with cover, especially dry grassland and shrub steppes with stone outcrops and stony semi-deserts, from 450m to 5,073m. Pallas's Cats are vulnerable to predation on open ground and show a strong preference for areas with cover such as ravines, rocky hillslopes and vegetated valleys. They avoid very open habitat such as extensive short grassland and lowland sandy desert basins found throughout the Tibetan Plateau; they are sometimes found deep within such open habitats along seasonal river courses. Although Pallas's Cats are well adapted for extreme cold, they avoid areas with deep snow and their distribution stops where prolonged snow cover of 15–20cm begins.

Pallas's Cat colouration is suited for blending into open habitats with rocky cover, a defence against faster, sight-hunting predators such as large raptors and canids (C).

Feeding ecology

Pallas's Cats hunt mainly small lagomorphs and rodents. Pikas are especially important prey across the range, typically comprising more than 50 per cent of the diet; gerbils, voles, hamsters, ground squirrels and young marmots are also commonly eaten. Following small mammals, small passerine birds are the most important category of prey. Occasional prey includes hares, hedgehogs, larger birds (including a record of a large raptor from Mongolia), lizards and invertebrates. There is one record of predation on a newborn Argali sheep lamb from the Ikh Nartiin Chuluu Nature Reserve, Mongolia. Depredation on domestic animals must be very rare if it occurs at all; poultry is largely absent from most of the species' range and local people do not report them taking young hoofstock.

Pallas's Cats may be active at any time but foraging is mostly crepuscular, which maximises the overlap between high activity by prey and low activity by predators (especially diurnal eagles) or competitors. They hunt by three distinct techniques: 'stalking' by creeping very slowly and stealthily around cover to locate and move close to prey; 'moving and flushing', used mainly in spring and summer by walking quickly or trotting among long grass and undergrowth to flush rodents and small birds; and waiting in ambush at rodent burrows, a technique used especially in winter. They are recorded scavenging from carcasses, including those of dead livestock. They are not known to cache food but they routinely take kills into dens and burrows to consume them.

Social and spatial behaviour

Pallas's Cat has been well studied only in Hustain Nuruu National Park, central Mongolia, where 29 cats were radio-collared; there are ongoing telemetry studies in Qinghai, China, and Daursky State Nature Reserve, Russia, with small numbers of cats collared. Pallas's Cat is solitary. Both sexes maintain enduring home ranges, with large, overlapping male ranges encompassing one to four female ranges which overlap much less with each other. It is likely they are territorial, at least in the breeding season; males during this period often have injuries consistent with fighting, presumably for access to females. Pallas's Cat ranges are very large for their size. Female territories range between 7.4km^2 and 125.2km^2, averaging 23.1km^2, compared to male territories of 21–207km^2, averaging 98.8km^2 (Hustain Nuruu National Park, Mongolia). Some resident Pallas's Cat adults in central Mongolia emigrate from their territories to establish new ranges, apparently unrelated to prey availability or intraspecific competition. Emigration occurs in August to October coinciding with peak numbers of Red Fox, Corsac Fox and various raptors following spring and summer breeding, which perhaps drives cats to seek new ranges. Adults are recorded emigrating a straight-line distance of 18–52km. One emigrating male made an exploratory, looping excursion measuring 170km over two months, covering an area of 1,040km^2.

There are few rigorous density estimates for Pallas's Cat, in part because they appear to occur at very low densities everywhere and are very difficult to survey. The most rigorous estimate is four to six Pallas's Cats per 100km^2 (Hustain Nuruu National Park, Mongolia). Winter snow-tracking in southern Russia produced extremely high estimates of 12–21.8 cats per 100km^2, but this is likely to be a poor technique to estimate density for this species and estimates should be treated with caution.

Reproduction and demography

Pallas's Cats live in habitats with severe environmental extremes and reproduction in the wild is highly seasonal. Mating occurs December–March and litters are born from the end of March through May. Oestrus (in captivity) is very short at 24–48 hours. Gestation lasts 66–75 days. Litter size averages three to four and exceptionally reaches as many as eight (captivity). Kittens reach independence at four to five months. Sexual maturity is 9–10 months for both sexes (captivity). Three female kittens in Mongolia dispersed 5–12km and became resident outside their natal ranges at 6.5–8 months old. All of them successfully bred for the first time at 10 months.

Mortality 68 per cent of kittens do not survive to disperse and adults have a 50 per cent chance of surviving until age three (Hustain Nuruu National Park). Most mortality occurs in winter between January and April. Predation is the main natural cause of mortality; large raptors (six occasions) and Red Fox (one occasion) accounted for 42 per cent of known deaths in Hustain Nuruu National Park. Predation by domestic dogs and killing by people accounted for a further 53 per cent of deaths. Captive Pallas's Cats, especially kittens, are highly susceptible to toxomoplasmosis, which has also been recorded rarely in the wild, though there is no evidence of widespread mortality.

Lifespan May be up to six years in the wild but is half of this on average; captive Pallas's cats can survive up to 11.5 years.

When caught in the open, the Pallas's Cat's main defence is to flatten itself to the ground, relying on immobility and camouflage to avoid detection.

STATUS AND THREATS

Pallas's Cats live in remote areas with low human densities but the species is nowhere common, is dependent on specific habitats and prey, and is easily killed on open ground, all of which makes it naturally susceptible to threats. The spread and intensification of livestock, agriculture and mining in Mongolia, China and Russia threaten large parts of Pallas's Cat range. Additionally, they suffer from the loss of prey as a result of extensive state-sanctioned poisoning campaigns to control populations of pikas and marmots, considered to be vectors of disease and competitors with livestock for grazing across central Asia and China. Pallas's Cat was widely hunted for its fur in the past and while international trade largely ceased in the 1980s, it is still hunted locally, especially in Mongolia where hunting is legal for domestic use. Despite bans on international trade, Pallas's Cat furs from Mongolia are openly sold into China. Close to settlements and nomads' camps, the species is often killed in traps intended for wolves and foxes, and domestic dogs constitute a key predator.

CITES Appendix II. Red List: Near Threatened. Population trend: Decreasing.

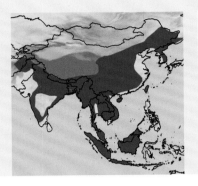

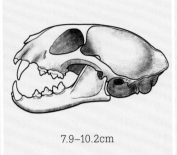

7.9–10.2cm

IUCN RED LIST (2008):
- Least Concern (Global)
- Vulnerable (Philippines)
- Critically Endangered (Iriomotejima)

Head-body length ♀ 38.8–65.5cm,
♂ 43.0–75.0cm
Tail 17.2–31.5cm
Weight ♀ 0.55–4.5kg, ♂ 0.74–7.1kg

Leopard Cat

Prionailurus bengalensis (Kerr, 1792)

Amur Leopard Cat

Iriomote Cat

Southern Leopard Cat

Taxonomy and phylogeny

The Leopard Cat is classified in the *Prionailurus* lineage where its closest relatives are the Fishing Cat and, more distantly, the Flat-headed Cat. Up to 12 subspecies are described including as many as seven island subspecies, but the taxonomy is in need of review. Limited genetic data suggest two highly variable subspecies on mainland Asia, divided into a northern subspecies, the Amur Leopard Cat *P. b. euptilurus* of the Russian Far East, north-eastern China, Korean Peninsula, Taiwan, and Japan's Tsushima and Iriomote Islands; and a southern subspecies, *P. b. bengalensis*, occurring everywhere else on the mainland (excluding possibly the Malay Peninsula; see below). Some populations were formerly classified as separate species, now considered invalid, notably the Iriomote Cat on Japan's Iriomote Island and the Amur Leopard Cat, both of which are now classified as *P. b. euptilurus*. There is recent genetic evidence that Leopard Cat populations south of the Kra Isthmus – on the Malay Peninsula, Borneo, Sumatra, Java and Bali – actually represent a separate Leopard Cat species (as for the now accepted separation of mainland and Sunda Clouded Leopards as two species, p184). Further analysis of a wider sample will help to resolve this issue, including the classification of the Leopard Cat in the Philippines, which is usually considered the subspecies Visayan Leopard Cat *P. b. rabori*. Leopard Cats readily hybridise with domestic cats which is apparently reported from the wild in rural areas, though there is no physical evidence of wild hybrids.

Description

The Leopard Cat varies considerably in size and appearance, depending on the region and on the season in northern populations. The smallest adults are from tropical islands and can weigh less than 1kg compared to northern individuals that may exceed 7kg; there are exceptional records from Russia of 8.2–9.9kg during late summer to autumn when Leopard Cats are hyperphagic before the severe winter. Colouration is extremely variable. Individuals in mainland tropical Asia tend to be richly coloured with yellow to tawny-brown or ginger-brown fur, and bold markings varying from large solid dabs to rosettes and blotches with dark tawny edges or centres. Island forms, including those on Borneo and Sumatra, tend to be less strongly marked with small, discrete spots and dabs on a background colour that varies widely from drab ginger-brown to very dark brown. Iriomote Cats are very dark; some individuals are blackish-grey with indistinct markings except on the face and underparts. At the other extreme, Amur Leopard Cats in temperate Russia, Korea and China are very pale ginger-grey to silver-grey in winter with long, dense fur that moults to a darker, summer coat of russet-brown to grey-brown. Complete melanism does not occur although there are occasional records of pseudomelanistic individuals with extensive enlarging and coalescing of the dark markings.

Similar species Dark forms of the Leopard Cat are similar to the closely related Fishing Cat, which is considerably larger and more heavily built; very young kittens of both species can

An Amur Leopard Cat hunts birds in reeds at the edge of a frozen lake, Taean region, South Korea.

be indistinguishable. Leopard Cats might also be confused with the spotted form of the Asiatic Golden Cat, which is much larger with a long, more tapering tail.

Distribution and habitat

The Leopard Cat is the most widespread of all small Asian felids. It is found in tropical, subtropical and temperate Asia from the Russian Far East, north-eastern China and the Korean Peninsula, through eastern China to the Tibetan Plateau and in the Himalayan foothills as far west as central Afghanistan; from northern Pakistan, Nepal and Bhutan to southern India; and throughout South-east Asia, including on Borneo, Java and Sumatra. The Leopard Cat occupies more islands than any other

felid including Hainan, Taiwan, the Philippine islands of Palawan, Panay, Negros and Cebu, and the Japanese islands of Iriomote and Tsushima; it is the only felid native to Japan. It is absent naturally from Sri Lanka.

The Leopard Cat occurs in a very wide variety of habitats with cover including all forest types ranging from lowland tropical rainforest to dry broadleaf and coniferous forest in the Himalayan foothills as high as 3,254m (eastern Nepal). They inhabit vegetated valleys in cold, temperate forest with winter snowfall in the northern range but they are limited to areas with shallow snow. They inhabit all kinds of woodland, scrub habitat, shrublands, marshes, wetlands and mangroves. They occur in wooded grasslands and shrub-grassland mosaics but they largely avoid open grasslands, steppes and rocky areas lacking vegetation. Leopard Cats are tolerant of human-modified habitats with cover including logged forest, farmlands such as sugar cane fields, and plantations of oil palm, coffee, rubber trees and tea. Leopard Cats may reach high densities in some open modified habitats that sustain high rodent numbers. They occur very close to human habitation, including in patches of suitable habitat in major metropolises, for example Miyun Reservoir and Yeyahu Nature Reserve in Beijing.

Feeding ecology

The Leopard Cat's diet is made up primarily of very small prey, chiefly small vertebrates. In most well-studied Leopard Cat populations, the most important prey category is made up of various mice and rats supplemented with other small mammals including squirrels, chipmunks, tree shrews, shrews and moles. Larger mammals recorded in the diet include hares, langurs, Lesser Mouse Deer and Wild Boar, most of which were probably scavenged. In Russia, it is reported to attack neonate ungulates including Roe Deer, Sika and Long-tailed Goral provided they are under a week old and when unguarded by the mother. Leopard Cats also eat birds up to the size of pheasants, as well as herptiles

and invertebrates. Three species of skinks, four snake species, a frog and a large cricket species are important prey to Iriomote Cats where rodents probably did not occur until humans introduced the Black Rat (which is now the most important prey by biomass). Leopard Cats are harmless to hoofstock but they prey on domestic poultry and are apparently easily captured with chickens as bait.

The Leopard Cat is a very active and versatile hunter that forages mainly on the ground and in low vegetation. However, it is an excellent, lightweight climber that is very much at home among slender branches and palm fronds where, although direct observations are few, hunting is bound to also occur. Similar to other *Prionailurus* felids, they have a strong affinity for water and are strong swimmers; they readily forage in the shallows for amphibians, freshwater crabs and other invertebrates. Hunting activity is variable, ranging from being strictly nocturnal at some sites to cathemeral, probably depending on prey availability and the presence of larger carnivores or humans. They readily scavenge, including reportedly inside caves where they eat dead or fallen bats and Cave Swiftlets. The cat also caches large kills; for example, a Ring-necked Pheasant was hidden in scrub and eaten by an Amur Leopard Cat over a number of sittings.

Social and spatial behaviour

The Leopard Cat follows the basic felid socio-spatial pattern. They are largely solitary with enduring ranges in which male ranges generally overlap one

Left: **The Leopard is a known predator of Leopard Cats. The effects of predation on Leopard Cat populations are poorly known except where heavy human predation, in the form of fur harvest and persecution, almost certainly produces declines.**

Below: **The Leopard Cat is often the most abundant felid where it occurs and the species most likely to be encountered by people (C).**

Leopard Cats are very agile climbers that seek refuge in trees when threatened or pursued by predators (C).

or more, smaller female ranges. Range size differs little between sexes in some populations, for example in Phu Khieo Wildlife Sanctuary, Thailand. Overlap between same-sex adults is considerable at the range edges and is usually minimal in exclusive, core areas, yet core areas overlap significantly in Phu Khieo. Range size is 1.4–37.1km^2 (females) and 2.8–28.9km^2 (males). Leopard Cats urine-mark and scrape in typical felid fashion but the degree of territorial defence is unclear. Iriomote Cat adults were aggressive to each other at experimental feeding stations but this may have been heightened by the situation's artificiality. Density estimates include 17–22 Leopard Cats per 100km^2 (subtropical and temperate Himalayan forest, Khangchendzonga Biosphere Reserve, India), 34 Leopard Cats per 100km^2 (Iriomotejima) and 37.5 Leopard Cats per 100km^2 (lowland rainforest and adjacent oil-palm plantations, Tabin Wildlife Reserve, Sabah).

Reproduction and demography

This is relatively poorly known from the wild but available information indicates that the species breeds aseasonally in most of its range, becoming more seasonal in temperate areas. The Amur Leopard Cat is apparently highly seasonal with births restricted to late February–May. Captive Leopard Cats are able to have two litters a year, though a single litter is probably typical in the wild. Gestation lasts 60–70 days. Litter size is one to four kittens, typically two to three. Sexual maturity occurs at 8–12 months (captivity); the earliest breeding in a captive female was at 13 months.

Mortality Estimated annual adult mortality varies from 8 per cent in a remote sanctuary with low human pressure (Phu Khieo Wildlife Sanctuary, Thailand) to 47 per cent in an accessible protected area (Khao Yai National Park, Thailand). Mortality is likely even higher outside protected areas close to people. Leopard Cats are probably killed occasionally by a wide variety of natural predators given their small size but there are few records; Leopards and domestic dogs are confirmed predators and Wild Boar are suspected predators of kittens on Tsushima. Humans are the main source of mortality for many populations, for example in much of South-east Asia and China. Roadkills are the major factor on Iriomotejima (at least 40 deaths from 1982 to 2006) and Tsushima (at least 43 deaths from 1992 to 2006).

Lifespan Unknown in the wild, up to 15 years in captivity.

STATUS AND THREATS

The Leopard Cat is widespread, adaptable and is able to live near people where tolerated. It reaches high densities in suitable habitat (including some anthropogenic landscapes) and is the most abundant felid in most of its range. However, it is the only Asian cat legally harvested for fur, for which it is heavily hunted in its temperate range where densities are naturally lower than elsewhere in the range. In 1985–1988, as many as 400,000 Leopard Cats were killed annually in China for the fur trade. China suspended its international trade in 1993, when a stockpile of a staggering 800,000 skins was estimated. The cessation of international trade has apparently reduced hunting pressure, though Leopard Cats are still legally hunted in China outside protected areas and skins are very common in Chinese fur warehouses. The impacts of such high harvests are unknown. Although hunting is illegal in subtropical and tropical Asia, the species is nevertheless widely killed for fur and meat, and in retaliation for poultry predation. The species is also targeted for the pet trade and is frequently sold in wildlife markets, for example in Jakarta, Java and in Vietnam. Many island populations are small and threatened by rapid development. Populations on Tsushima and the Philippine islands of Panay, Negros and Cebu are declining. The Iriomotejima population numbers around 100 and is well protected in central montane habitats, but is declining in coastal lowlands.

CITES Appendix I (Bangladesh, India and Thailand), Appendix II elsewhere. Red List: Least Concern (global), Critically Endangered (Iriomotejima), Vulnerable (Philippines). Population trend: Stable (global), decreasing on some islands.

9–9.8cm

IUCN RED LIST (2015): Endangered

Head-body length
♀ 44.6–52.1cm, ♂ 41–61cm
Tail 12.8–16.9cm
Weight ♀ 1.5–1.9kg, ♂ 1.5–2.2kg

Flat-headed Cat

Prionailurus planiceps (Vigors & Horsfield, 1827)

Taxonomy and phylogeny

The Flat-headed Cat is classified in the *Prionailurus* lineage and is thought to be most closely related to the Leopard Cat and Fishing Cat. No subspecies have been described.

Description

The Flat-headed Cat is a small, very distinctive felid with a short, tubular body, relatively short, slender legs and a stubby tail. The head is small with a compact, foreshortened face and closely set, large eyes, a flattened forehead and small rounded ears.

The feet are partially webbed and the claw sheaths are reduced so that the claws protrude visibly. This has led to the claws being described as non-retractile, though they have the typical felid ability to protract their claws. Flat-headed Cats are dark roan-brown graduating to rich rusty-brown on the head. The face has bright white on the cheeks, eyebrows and under the eyes, and contrasting dark rusty-brown cheek and eyebrow stripes. The body fur is dense and soft, and is largely unmarked except for light dappling and banding on the legs and belly; the tail is sometimes faintly banded.

Similar species Very distinct from all other cats. It broadly resembles the Rusty-spotted Cat but the two species are separated by range. It could be confused for a very small, stocky domestic cat at a glance.

Distribution and habitat

The Flat-headed Cat has a limited range restricted to Borneo, Sumatra and Peninsular Malaysia. It may occur in extreme southern Thailand on the Thai-Malaysia border where there are records from Pru Toh Daeng Peat Swamp Forest but none since 1995. Two kittens confiscated from Thai wildlife traffickers in 2005 were believed to have been smuggled from Peninsular Malaysia for the pet trade. Flat-headed Cats are closely associated with moist, lowland forested habitats and wetlands. More than 80 per cent of historical and recent records of the species occur below 100m above sea level, and more than 70 per cent of records are within 3km of large rivers and water sources. They inhabit primary and secondary forest, peat-swamp forest, mangrove and coastal scrub-forest. Although there are reports from secondary forest and oil-palm plantations suggesting they are tolerant of some habitat modification, records from altered habitats are questionable or very few.

Feeding ecology

The Flat-headed Cat is one of world's least known felids and its ecology in the wild is largely a mystery.

A Flat-headed Cat hunting frogs in sodden vegetation at the edges of the Menanggul River, Sabah, Borneo.

The species' unique morphology, behaviour and habitat preferences suggest it is adapted to forage for aquatic prey in shallow water and along muddy riverbanks. As well as the modifications to the forepaws, Flat-headed Cat teeth are very sharp and oversized (the first and second upper premolars) and are considered to be adaptations for grasping slippery prey. Captive animals are attracted to water, readily submerging themselves and feeling for food in pools with spread paws, in a technique similar to raccoons. The stomach contents of a handful of dead animals examined in the wild contained fish and crustaceans. Captives quickly dispatch mice and rats with a nape bite, and wild individuals almost certainly take small mammals, reptiles and amphibians. They are sometimes killed in traps set at poultry coops suggesting they occasionally kill domestic fowl. Most records, particularly camera-trap images, are at night suggesting that they are largely nocturnal.

Social and spatial behaviour

Flat-headed Cats have never been radio-collared and their socio-spatial behaviour is unknown. Camera-trap images are almost always of single animals; like most small felids, they are probably essentially solitary with enduring and semi-exclusive ranges, though everything about these patterns remains unconfirmed. Density is unknown. Of all South-east Asian felids, they are the least frequently photographed in camera-trap surveys by a wide margin; only 17 camera-trap photographs of the species existed by 2009 compared to many hundreds or thousands for all other sympatric felids. This suggests very low densities, though cameras are usually placed to maximise photographs of large felids and rarely target high-quality Flat-headed Cat habitats, such as along the banks of water sources. It is therefore possible they are more common than suggested by camera-trapping.

Reproduction and demography

Completely unknown from the wild. Gestation is 56 days (captivity) and females have one to two kittens based on only three captive litters.

Mortality Unknown, but being so small they are presumably vulnerable to a wide array of predators.

Lifespan Unknown in the wild, up to 14 years in captivity.

Flat-headed Cats are very rarely observed hunting in the wild. They are known to use their partially webbed feet to feel for prey in shallow water and muddy edges.

STATUS AND THREATS

Flat-headed Cats are known only from 107 (as of 2009) physical records and sightings. This is likely due to a combination of rarity and sampling bias, but even with markedly increased survey effort in the last decade, there are only a handful of sites where the species is repeatedly recorded, for example Deramakot Forest Reserve and Kinabatangan Wildlife Sanctuary, Sabah. The Flat-headed Cat's restricted distribution and very close association with moist, forested habitats is gravely concerning given the rapid pace of habitat loss. As of 2009, an estimated 54–68 per cent of suitable Flat-headed Cat habitat had been converted by people, especially clearing and draining of forest wetlands for croplands and forestry. Overfishing and freshwater pollution from agriculture and mining are likely to exacerbate declines driven by habitat loss, and hunting by people may exert strong local effects. Flat-headed Cat skins often occur in longhouses in Sarawak. Live animals, usually kittens, occasionally appear in the pet trade. Many authorities now consider the Flat-headed Cat to be South-east Asia's most threatened small felid.

CITES Appendix I. Red List: Endangered. Population trend: Decreasing.

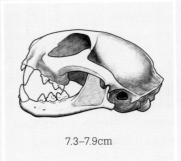

7.3–7.9cm

● **IUCN RED LIST (2008):**
Vulnerable

Head-body length 35–48cm
Tail 15–29.8cm
Weight ♀ 1.0–1.1kg, ♂ 1.5–1.6kg

Rusty-spotted Cat

Prionailurus rubiginosus

(I. Geoffroy Saint-Hilaire, 1831)

Taxonomy and phylogeny

The Rusty-spotted Cat is classified in the *Prionailurus* lineage. It is thought to have diverged as an early offshoot within this lineage and thus may be less closely related to the other three *Prionailurus* species than they are to each other. Three subspecies are recognised, two from Sri Lanka representing lowland and upland populations, and one subspecies from India; none has yet been confirmed by molecular analysis. A single observation from Ruhuna National Park (Sri Lanka) indicates that wild Rusty-spotted Cats sometimes mate with domestic cats, though it is unknown if hybridisation occurs.

Description

The Rusty-spotted Cat is one of the smallest felids, like a diminutive domestic cat at first glance, and similar in weight but with slightly different proportions to the Black-footed Cat. The short, smooth fur is rufous-brown or greyish-brown, with rows of rust-red to dark brown spots that sometimes form complete stripes on the nape, shoulders and upper flanks. The underparts are white or pale cream. The tail is relatively thick and tubular, often faintly banded and with a darkish tip.

Similar species The Rusty-spotted Cat is sympatric with the closely related Leopard Cat in parts of its Indian range but is smaller and not as boldly marked. It is easily mistaken for a very small domestic cat.

Distribution and habitat

The Rusty-spotted Cat is endemic to India and Sri Lanka. It has recently been confirmed from Pilibhit Territorial Forest Division and Katarniaghat Wildlife Sanctuary, northern India on the border with Nepal and is strongly suspected to occur in Bardia National Park, Nepal. The species was long regarded as a forest specialist but it is now known to have a broad habitat tolerance, occurring in all kinds of moist and dry forest, bamboo forest, wooded grasslands, arid shrublands, scrublands and vegetated, rocky habitats. They appear to be mostly absent from evergreen forest but they inhabit humid montane forest to 2,100m in Sri Lanka. Rusty-spotted

This Rusty-spotted Cat warming itself in the morning sun has chosen an elevated vantage point from which it can also search for prey, Ranthambore National Park, India.

Cats tolerate modified habitats provided there is cover, including agriculture, tea plantations and forest plantations which are often rich in rodent and amphibian prey; they hunt in seasonally inundated fields, such as maize and rice, which have abundant frogs and toads. They sometimes live very close to people, even in abandoned dwellings in inhabited villages.

Feeding ecology

Rusty-spotted Cats have a reputation for being especially fierce and taking very large prey but this is based apparently on observations of tame individuals in unnatural settings. For example, an eight-month-old pet Rusty-spotted Cat almost succeeded in killing a tame gazelle lamb several times its size by a throttling throat bite before people intervened, but such attacks in the wild are highly unlikely. The known diet comprises small rodents such as Indian Gerbil, bandicoot rats and various mouse species, especially the Little Indian Field Mouse, as well as small birds, hatchlings, reptiles, toads and invertebrates. It sometimes kills free-

Right: **This captive Rusty-spotted Cat shows the species' contrasting facial stripes, often the most conspicuous markings compared to the lightly marked body (C).**

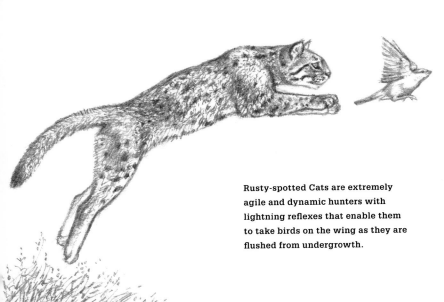

Rusty-spotted Cats are extremely agile and dynamic hunters with lightning reflexes that enable them to take birds on the wing as they are flushed from undergrowth.

ranging domestic poultry, mainly chicks; it rarely if ever enters coops or dwellings for poultry. Most sightings of foraging cats indicate they are largely nocturnal and hunt on the ground but they are superb climbers and possibly also hunt arboreally. They often adopt a sit-and-wait hunting technique, positioning themselves on a branch, rock or other elevated point and listening for the sounds of small prey on the ground. There is one report of a hunting cat launching itself from the low branches of a tree at prey on the ground below. A chase of 50m is recorded, resulting in a shrew kill. A handful of direct observations of successful hunts demonstrate that smaller prey, e.g. Indian Gerbils, are killed with a nape bite whereas larger prey, e.g. bandicoot rats, are killed by asphyxiation with a throat bite.

Social and spatial behaviour

The Rusty-spotted Cat has never been intensively studied or radio-collared, but available information suggests a typical small felid socio-spatial system.

Camera-trap photos of the same individuals in localised areas indicate enduring home ranges. Archetypal territorial patrolling behaviour by a Sri Lankan male included spraying low-hanging branches and bushes, and rubbing vegetation with the cheeks and chest. There is no information on range sizes or density.

Reproduction and demography

Unknown from the wild; in captivity, reproduction is aseasonal. Two wild litters, one each from India and Sri Lanka, were both found in the month of February. Gestation is typically 67–71 days, with a range of 66–79 days. Litter size is between one and three kittens.

Mortality Unknown but their tiny size suggests they are vulnerable to predation by a wide variety of larger carnivores including domestic dogs, and perhaps owls and pythons. A mother with two young kittens was observed to lose one to an Indian Cobra (Sri Lanka). Rusty-spotted Cats are known to flee into trees when threatened.

Lifespan Unkown in the wild, up to 12 years in captivity.

A female Rusty-spotted Cat in Sri Lanka fetches its remaining kitten moments after an Indian Cobra killed her other kitten.

STATUS AND THREATS

The Rusty-spotted Cat is regarded as rare. Surveys over the last decade suggest it is more widespread than previously thought, though even the new data show that it is not common anywhere. They are found fairly often in association with humans and they can prosper in human-modified habitats given their tiny size and value in controlling rodents, provided tolerance is shown by people. The very widespread use of insecticides and rodenticides in much of the range is cause for grave concern, though impacts are unknown. Rusty-spotted Cats are often killed on roads and occasionally by domestic dogs and cats. In Sri Lanka, rural people have killed Rusty-spotted Cats claiming to mistake them for young Leopards, which are feared. They are sometimes killed as bycatch or intentionally as a perceived threat to poultry, though the species is very rarely to blame. Rural people in western and central India do not regard the Rusty-spotted Cat as a predator of poultry.

CITES Appendix I (India), Appendix II (Sri Lanka). Red List: Vulnerable. Population trend: Decreasing.

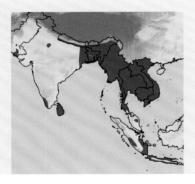

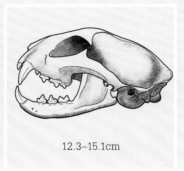

12.3–15.1cm

● **IUCN RED LIST (2008):**
Endangered

Head-body length ♀ 57–74.3cm,
♂ 66–115cm
Tail 24–40cm
Weight ♀ 5.1–6.8kg, ♂ 8.5–16.0kg

Fishing Cat

Prionailurus viverrinus (Bennett, 1833)

Taxonomy and phylogeny

The Fishing Cat is classified in the *Prionailurus* lineage where its closest relatives are the Leopard Cat and, more distantly, the Flat-headed Cat. Two subspecies are recognised, one from Java and one from the rest of the range, i.e. mainland Asia and Sri Lanka. These are based on archaic descriptions without strong evidence and Javan individuals cannot be distinguished morphologically from mainland animals. Recent but limited genetic data indicate a weak difference between populations on either side of the Kra Isthmus but sampling was insufficient to make a clear distinction.

Description

The Fishing Cat is by far the largest of the *Prionailurus* cats, with a muscular, robust body, stocky legs and a relatively short, well-muscled tail. The head is blocky and powerfully built, with small, rounded ears that are black-backed with a white central spot. Fishing Cat feet are partially webbed and the large claws protrude partially from the claw sheaths; although the claws may appear extended, they are fully protractile. The body fur is typically olive-grey, sometimes with a slate-grey or russet tinge, fading to pale underparts. The body is covered in dark brown to black spots that typically coalesce into long blotches or stripes on the nape, shoulders and back. Albinism is recorded from the Hail Haor Wetland, north-eastern Bangladesh.

Similar species The Fishing Cat is likely to be confused only with dark forms of the closely related Leopard Cat but the latter is considerably smaller and more gracile; very young kittens of both species can be indistinguishable.

Distribution and habitat

The Fishing Cat has a relatively wide but extremely fragmented distribution across South and South-east Asia. The main strongholds are the *terai* lowlands across southern Nepal and northern India, north-eastern and eastern India, north-eastern Bangladesh and the island of Sri Lanka where it appears to be widely distributed. It is thought to be recently extinct in many areas of former occurrence, including the Indus Valley, Pakistan, western India and possibly north-western India where its presence in the former stronghold, Keoladeo Ghana National Park, Rajasthan, is now uncertain. In South-east Asia, the distribution is extremely patchy and reduced, with known populations restricted to a few sites each in Thailand and Java, Indonesia, and very few recent records from Cambodia, Laos, Burma and Vietnam. Reports from Peninsular Malaysia, Sumatra, Taiwan and south-western China are erroneous or equivocal.

Fishing Cats are strongly associated with wetland habitats including marshes, reedbeds, dense *terai* grasslands (Nepal), riverine woodlands, coastal wetlands and mangroves. They occur in evergreen and dry forest closely associated with well-watered areas such as marshland, oxbow lakes and slow-moving rivers. They are sometimes found

There are very few observations of Fishing Cats hunting in the wild. Much of their behaviour is inferred from captives in naturalistic enclosures with opportunities to forage, such as this individual in Singapore Zoo (C)

in anthropogenic habitats including aquaculture ponds, rice paddies and along canals near major cities (for example, Kolkata, India, and Colombo, Sri Lanka) but they cannot tolerate extreme wetland modification that is typical for the region. Fishing Cats usually occur from sea level to 1,000m; there is one record from 1,525m in the Indian Himalayan foothills.

Feeding ecology

The Fishing Cat has a diet dominated by aquatic prey, particularly fish as well as crustaceans, molluscs, amphibians, water-associated reptiles, including Bengal Monitors and snakes, and semi-aquatic rodents and those living in wetland habitats. Small rodents occurred in 70 per cent of scats of Fishing Cats collected in Keoladeo Ghana National Park, India. While the Fishing Cat's forefeet are somewhat adapted for catching slippery prey, it otherwise has few adaptations for a specialised diet; its dentition is robust and typical of a more generalised felid diet. Larger mammals recorded as occasional prey include hares, Small Indian Civets and Chital neonates. Birds, particularly ducks, coots and shorebirds, are also taken, including apparently from the water. Insects are common in scats but are unlikely to contribute much to energetic requirements. They sometimes kill poultry. They are often blamed for killing juvenile goats and very young calves, and yet while the Fishing Cat is sufficiently powerful to kill small stock, there are very few reports with unequivocal evidence. Reports of them killing human infants are unsubstantiated and extremely unlikely.

The Fishing Cat is a very capable swimmer that appears to hunt mainly by a sit-and-wait technique at the water's edge to locate prey. They readily enter the water in pursuit of prey, actively hunting in the shallows and fully submerging themselves while swimming after fish. Foraging is thought to be nocturnal, based on camera-trap records and telemetry data, but these are very few and are biased towards areas with high human activity where cats are more likely to avoid diurnalism. Fishing Cats are known to scavenge from dead livestock, which is likely to have fuelled their reputation as stock killers, as well as from the kills of larger carnivores including Tigers.

Social and spatial behaviour

The Fishing Cat is very poorly known. The species has been radio-collared only in one small study in Nepal and during a larger, ongoing study in Khao Sam Roi Yot National Park, Thailand, which is likely to produce some more comprehensive data – 17 cats

have been collared (by October 2014) but the information is yet to be published. Limited available information indicates a typical small felid solitary socio-spatial system with small female ranges overlapped by larger male rages. Range sizes from limited monitoring in *terai* grasslands are 4–6km^2 (two females) and 22km^2 (one male; Chitwan National Park, Nepal). There are no rigorous density estimates.

Reproduction and demography

This is poorly known from the wild. In captivity, gestation lasts 63–70 days and litter size is usually one to three kittens (exceptionally, four); the mean from 13 captive litters was 2.6 kittens. There is little evidence for seasonal breeding, although this is often assumed; even with very few records, kittens have been recorded from the wild in January–June, suggesting either weak seasonality or merely limits in sampling. Sexual maturity in one captive female was 15 months.

Mortality There are no records of natural mortality; given their habitat preferences, large crocodiles and pythons are potentially major predators. Humans and domestic dogs are the main cause of mortality where monitored.

Lifespan Unknown from the wild and up to 12 years in captivity.

STATUS AND THREATS

Until recently, Fishing Cats were considered widespread and relatively common, but extremely rapid conversion of wetlands, floodplains and mangroves for human settlement and agriculture throughout tropical Asia has prompted a severe decline in most of the range. The relatively recent introduction of aquaculture and prawn farms into mangrove and coastal habitats is a particularly pervasive threat. Suitable remaining wetland habitat is further threatened by overfishing and pollution, which is exacerbated by human persecution of Fishing Cats. They are killed mainly as perceived poultry or livestock predators, by fishermen who believe they take fish from nets, as bycatch in snares set for other species or, rarely, as food in some areas. Given the high human densities associated with most wetland habitat in their range, there are very few places where direct and indirect anthropogenic threats are not prevalent. The Fishing Cat's strongest conservation prospects probably lie in the lowlands south of the Himalayas where it mostly occurs in protected areas, in Sri Lanka, perhaps in the Sundarbans (Bangladesh and West Bengal, India) and perhaps in a handful of coastal sites in Thailand. Its prospects are poor in most of South-east Asia where it is likely endangered or extirpated. The isolated Javan population is probably Critically Endangered.

Cites Appendix II. Red List: Endangered. Population trend: Decreasing.

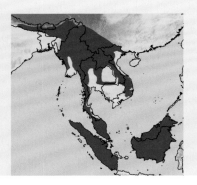

8.8–10.3cm

● **IUCN RED LIST (2008):**
Vulnerable

Head-body length 45–62cm
Tail 35.6–53.5cm
Weight 2.5–5 kg

Marbled Cat

Pardofelis marmorata (Martin, 1837)

Taxonomy and phylogeny

The Marbled Cat was once considered a close relative of the Clouded Leopard based on superficial morphological similarities including the heavily blotched pelage, elongated tail, oversized feet and enlarged dentition. Genetic analyses show conclusively the species belongs in the *Pardofelis* lineage with the Bay Cat and Asiatic Golden Cat. These latter two species are more closely related to each other than either is to the Marbled Cat, which is classified in its own genus. There is increasing evidence that Marbled Cat populations on the mainland and on Sumatra and Borneo are actually two separate, closely related species (as for mainland and Sunda Clouded Leopards).

Description

The Marbled Cat is the size of a large, long-bodied domestic cat covered with dense, soft fur that gives an overall larger appearance and with an extremely long, thickly furred tail. The tail is very long, sometimes exceeding the head-body length and is a very distinctive feature in the field. When walking and relaxed, the Marbled Cat assumes a characteristic, slightly arched body posture with the tail held horizontally in a continuous, straight line from the body. It has a relatively small, rounded head with a broad, short face and rounded ears with a central white spot on the back. The paws are large and broad, likely reflecting the species' advanced arborealism. The background fur colour is variable, with various shades of grey-buff, yellow-brown or red-brown and is very richly patterned with large, dark-bordered blotches that graduate to small dabs on the legs. The tail is heavily patterned with large solid blotches along its length that sometimes form rings towards the tip. Melanism occurs rarely – there is one unambiguous camera-trap record from Bukit Barisan Selatan National Park, Sumatra.

Similar species The Marbled Cat closely resembles a small Clouded Leopard but is markedly smaller, and lacks the distinctive heavy-headed appearance of the Clouded Leopard and its more discrete blotches with bold edges.

Distribution and habitat

The Marbled Cat occurs in a narrow tropical band south of the Himalayas from eastern Nepal through northern India, Bhutan and south-west China, and patchily throughout Indochina from northern Burma to the Malaysian peninsula, Borneo and Sumatra. It may occur in extreme northern Bangladesh but there are no certain records. Marbled Cats are restricted to forested habitats, chiefly undisturbed evergreen, deciduous and tropical forest from sea level to 3,000m in the Himalayan foothills. They occur at low densities in disturbed, seasonally flooded peat forest (e.g. Sabangau Forest, Central Kalimantan, Indonesia) and are able to occupy secondary and logged forest, though it is unknown whether modified habitat is suboptimal. They are not recorded from heavily converted habitat such as oil-palm plantations.

Feeding ecology

The Marbled Cat is one of the least-studied felids and its ecology is very poorly known. Only one individual, an adult female, has been radio-collared and she was tracked only for one month in Phu Khieo Wildlife Sanctuary, Thailand. The Marbled Cat's morphology suggests a high degree of arborealism. They are highly agile climbers able to rapidly descend trees head-first, and there are brief sightings of them hunting in trees, including

Below: **A wild Marbled Cat in lowland rainforest, Tawau Hills National Park, Sabah, Malaysian Borneo.**

stalking birds, though no kills have been observed. Camera-trapping demonstrates they also move about on the ground and they presumably hunt both on the forest floor and in trees. The diet is likely to be dominated by small arboreal and terrestrial vertebrates, such as rodents, birds and herptiles; captive animals readily eat squirrels, rats, birds and frogs. The enlarged dentition, particularly of the canines, suggests it is equipped to overpower larger prey. A Marbled Cat in Thailand was disturbed in the act of killing a juvenile Phayre's Leaf Monkey estimated to weigh at least as much as the cat. It is unknown if Marbled Cats scavenge, but a captive individual refused carrion. Most camera-trap photographs are diurnal, though the sample is small, and the radio-collared female was active at night. Activity patterns are probably variable, depending on the presence of larger felids and people.

The Marbled Cat is a very adept climber that almost certainly takes arboreal prey including birds and perhaps primates during treetop hunts.

Social and spatial behaviour

Virtually unknown. Occasional sightings of adult pairs have fostered speculation that Marbled Cats form long-term pair bonds but most camera-trap photos are lone adults, suggesting a typical, solitary felid socio-spatial pattern. The collared Thai female used a range of 5.3km^2 in one month. There are no density estimates; Marbled Cats are generally rare in camera-trap surveys and in Asian wildlife markets, possibly a sign of naturally low densities. They may reach higher densities in Borneo than in mainland populations.

Reproduction and demography

Very poorly known with limited information available only from captivity. Gestation is 66–82 days, and litters average two kittens based on only two captive births. Captive females are sexually mature at 21–22 months.

Mortality Unknown. Potential predators include large cats and domestic dogs but there are no known records.

Lifespan Up to 12 years in captivity.

STATUS AND THREATS

Marbled Cats appear to be naturally rare and are forest-dependent, suggesting they are particularly vulnerable to habitat loss which is very prevalent throughout their range; South-east Asia has the world's fastest deforestation rate due to logging and conversion for human settlement and agriculture, including plantations especially of oil palm. The species is rare in wildlife markets but is killed opportunistically for fur and parts, which is potentially a serious threat especially in concert with habitat loss. They are occasionally killed as poultry pests. CITES Appendix I. Red List: Vulnerable. Population trend: Decreasing.

8.5–10cm

IUCN RED LIST (2008): Endangered
Head-body length 53.3–67cm
Tail 32.0–39.1cm
Weight 2kg (emaciated ♀)

Bay Cat

Catopuma badia (Gray, 1874)

Bornean Bay Cat

Grey form

Taxonomy and phylogeny

The Bay Cat is closely related to the Asiatic Golden Cat and was once considered a small, island subspecies of the latter. However, genetic analysis leaves no doubt that the two species are distinct, and they are classified as the only members of the genus *Catopuma*. They diverged from a common ancestor 4.9–5.3 million years ago, well before Borneo's isolation as an island. These two species and the closely related Marbled Cat comprise the *Pardofelis* lineage.

Description

The Bay Cat is about the size of a large, long-bodied domestic cat with a conspicuously long tail. It resembles a small, slender Asiatic Golden Cat with a

Red form

proportionally smaller, rounded head and stubby, rounded ears which are set rather low on the side of the head. It occurs in two forms, a rich rusty-red to mahogany-red, and a grey form with variable red undertones especially along the transition from the body colour to the paler underparts. There is some intergradation between forms – predominantly red colouration is more common among the small number of museum specimens but camera-trap samples indicate red and grey forms are probably equally common. Bay Cats are unmarked except for stripes on the forehead and cheeks, and faint spotting along the transition between the upper body colour and underparts. The back of the ear is black without a white spot. The bright white underside to the tail with a dark dorsal tip is distinctive in the field.

Similar species The Bay Cat is dissimilar to all other sympatric felids. The similar Asiatic Golden Cat does not occur on Borneo.

Distribution and habitat

The Bay Cat is endemic to Borneo. Historically, they have been closely associated with dense lowland

forest and riverine forest habitats under 800m but camera-trap surveys in the last decade have considerably expanded their known habitat tolerances. Two camera-trap records (2010) from 1,459–1,451m in Malaysian Borneo's Kelabit Highlands suggest they occur more widely than generally known in upland forest, but most tracts have never been surveyed. They do not appear to inhabit swampy regions of extreme lowlands; intensive surveys of the Lower Kinabatangan (Sabah) and the peat swamp forests of the Sabangau (Kalimantan) have not yielded any Bay Cats. Camera-trap surveys also demonstrate at least some ability to persist in human-modified forest, including records from recently logged secondary forest and disturbed forest-plantation mosaics. However, they have not been found during surveys of oil-palm plantations, and it appears that relatively dense forest is required for the species to persist.

Feeding ecology

The Bay Cat is one of the least-studied felids and its ecology is very poorly known. Small vertebrates presumably comprise most of the diet. Two Bay Cats were trapped in 2003 at an animal dealer's pheasant aviaries suggesting they might attack domestic poultry, though complaints are very rare. Camera-trap records occur around the clock with an apparent bias towards diurnality, perhaps to avoid Sunda Clouded Leopards and/or related to prey activity, for example of diurnal terrestrial birds that may comprise important prey. They are likely to forage mainly on the ground.

Social and spatial behaviour

Unknown. Most camera-trap photos are lone adult individuals suggesting a typical, solitary felid socio-spatial pattern. Bay Cats are rarely photographed during camera-trapping surveys, suggesting they occur at low densities; for example, Bay Cats were photographed 25 times at four sites in eastern Sabah over four years compared to 259 images of Sunda Clouded Leopards and more than 1,000 images of Leopard Cats. There is no information on home range size or density.

Reproduction and demography

There is no information on reproduction including from captivity; there are no captive Bay Cats at the time of writing and the species has never bred in captivity.

Mortality The species' potential predators include Sunda Clouded Leopard, perhaps large reptiles including Reticulated Python and Estuarine Crocodile, and domestic dogs. However there are no known records and predation is likely to be uncommon.

Lifespan Unknown.

Bay Cats are thought to hunt mainly on the ground where terrestrial-foraging birds such as partridges and pheasants are likely to be important prey.

STATUS AND THREATS

The Bay Cat's apparent rarity and its dependence on forest raise concerns for its conservation prospects. Forest conversion, especially to oil-palm plantations, is regarded as a serious threat. Their rarity and value are known to animal dealers, which elevates illegal-trapping pressure.

CITES Appendix II. Red List: Endangered. Population trend: Decreasing.

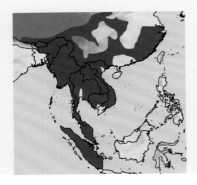

11.9–15.7cm

● **IUCN RED LIST (2008):**
Near Threatened

Head-body length ♀ 66–94cm,
♂ 75–105cm
Tail 42.5–58cm
Weight ♀ 8.5kg, ♂ 12–15.8kg

Asiatic Golden Cat

Catopuma temminckii

(Vigors & Horsfield, 1827)

Temminck's Golden Cat

tristis form

Taxonomy and phylogeny

The Asiatic Golden Cat is classified in the genus *Catopuma* with the closely related Bay Cat, which was once considered the same species, but this is invalidated by genetic analyses that demonstrate they are two distinct species with a common ancestor approximately 4.9–5.3 million years ago. Together with the Marbled Cat, they comprise the *Pardofelis* lineage. The Asiatic Golden Cat is not closely related to the African Golden Cat (see Description) with which it was once classified. Three subspecies of the Asiatic Golden Cat are usually described, largely on the basis of pelage differences which are unlikely to survive scrutiny by genetic analysis.

Description

The Asiatic Golden Cat is a medium-sized, robustly built cat with a relatively long, slender tail. The coat is highly variable, usually golden-brown to rich russet-brown but varying from pale tawny to dark coffee-brown and dark slate-grey. Uniformly coloured individuals are largely unmarked except for rich facial stripes and spotting on the chest, belly and inner limbs; some individuals are marked on the body with indistinct blotches against a pale background giving a faint marbled or 'watermarked' appearance. A richly spotted 'Ocelot' morph with light greyish fur and large russet blotches with dark borders has been recorded from Bhutan, China and Burma (this form is also known as *tristis*, reflecting its early classification as a separate but invalid subspecies). Completely melanistic individuals occur. Except in black individuals, the underside of the tail is conspicuously bright white with a dark upper tip.

Similar species The Bay Cat closely resembles a small Asiatic Golden Cat but the two species are not sympatric. The *tristis* form is similar to the Marbled Cat which is smaller with a distinctive very long, tubular tail. The African Golden Cat is very similar in appearance to the Asiatic Golden Cat but the two species are not closely related and are not sympatric.

Distribution and habitat

The Asiatic Golden Cat occurs along the southern slopes of the Himalayas from extreme eastern Nepal through southern Bhutan and north-east India, across southern and south-east China, south-east

Left: **In addition to completely black individuals, dark variants of the Asiatic Golden Cat include a deep slate-grey and a dark coffee-brown sometimes referred to as the 'cold-brown' form (C).**

Below: **A marbled or 'watermarked' Golden Cat with faint blotching, intermediate between the unicolour and *tristis* form. This form is relatively common in Endau Rompin National Park and surrounds, Malaysia where this individual was camera-trapped.**

habitats including thick scrubland, scrub-grasslands and upland mosaics of dwarf bamboo-grassland, and dwarf rhododendron-grassland. They sometimes occur in secondary and disturbed forest and are occasionally sighted or killed near human settlements, including in open agricultural areas, small forest fragments and plantations of oil palm and coffee. However, these records are very few and Golden Cats do not permanently occupy heavily modified habitats. Asiatic Golden Cats occur from sea level typically to 3,000m; there are higher records from the Himalayan slopes in Bhutan (3,738m) and Sikkim, India (3,960m).

Bangladesh, through most of South-east Asia and on Sumatra. They do not occur on Borneo. Asiatic Golden Cats are found primarily in forested habitats, including lowland and upland rainforest, dry deciduous forest, evergreen forest and montane forest. It is occasionally recorded in more open

Feeding ecology

The Asiatic Golden Cat is very poorly known; like most Asian small felids, it has never been the focus of a comprehensive research effort. Confirmed prey includes a variety of small to medium-sized

vertebrates. Rodents are likely a mainstay of the diet, with various mice, rats and squirrels recorded as prey. The largest confirmed prey taken by a Golden Cat is a Dusky Leaf Monkey (~6.5kg; Taman Negara National Park, Malaysia). In the same study, two species of mouse deer, Asiatic Brush-tailed Porcupine, birds, snakes and lizards were also eaten. Asiatic Golden Cats are powerfully built and are reputed to kill medium-sized ungulates including muntjacs and livestock to the size of very young cattle and domestic buffalo calves. The latter is unlikely; confirmed records of livestock kills are from people shooting them at carcasses from which they were probably scavenging. They sometimes raid poultry. Asiatic Golden Cats appear to be cathemeral; based on camera-trap photos and two radio-collared cats, they are somewhat more active by day and around dusk or dawn than at night in a variety of sites in Indonesia, Malaysia, Burma and Thailand. Foraging is thought to be mostly terrestrial, though they are very capable climbers and presumably take some prey arboreally.

Social and spatial behaviour

Only three Asiatic Golden Cats have been radio-collared, and data on two adults (from Phu Khieo Wildlife Sanctuary, Thailand) suggest a typical solitary felid socio-spatial system. The male's range ($47.7km^2$) was larger than the female's ($32.6km^2$) and they overlapped significantly; about half of the male's range overlapped 78 per cent of the female's range, suggesting male ranges encompass part or all of multiple female ranges. The same individuals travelled on average 900m–1.3km (dry season/wet season, female) and 2.1–3km (wet season/dry season, male), with a maximum daily distance of 3km (female) to 9.3km (male). There are no rigorous density estimates for the Asiatic Golden Cat; in 14 protected areas in Thailand, they were camera-trapped 81 times which was comparable to Clouded Leopards (79 photographs) and ranked fourth in relative abundance of six cat species, after Leopards, Tigers and Leopard Cats.

Reproduction and demography

This is unknown from the wild. In captivity, reproduction is aseasonal and gestation lasts 78–81 days. The litter size is typically one, rarely two and exceptionally three kittens; of 32 captive litters, 29 were singletons and three were twins. Sexual maturity in captive animals is 18–24 months; the oldest female to breed was 14.5 years old.

Mortality This is essentially unknown; Tigers, Leopards and possibly Clouded Leopards and Dholes are potential predators. People are probably a major source of mortality.

Lifespan Unknown in the wild, up to 17 years in captivity.

Tree-scratching has a dual purpose. As well as leaving territorial marks for conspecifics, it helps with maintaining the claws by removing broken splinters and shards. Claw fragments are sometimes found embedded in marked trees.

STATUS AND THREATS

Asiatic Golden Cats are threatened by forest loss and illegal hunting, which are widespread throughout their range, but their status and the degree of threat is poorly known. Despite numerous camera-trap surveys, records are very scarce in Bangladesh, Cambodia, China, India and Nepal. They are more widely distributed in Thailand, Bhutan, Indonesia (Sumatra), Lao PDR, Malaysia, Burma and Vietnam. Their largely terrestrial behaviour makes them vulnerable to snaring or hunting with dogs, and they are often targeted where larger mammals are extirpated, for example for meat in the Chittagong Hill Tracts, south-east Bangladesh. Skins are traded heavily in China and Burma where hunting pressure is regarded as high. They are occasionally killed as poultry pests. Cites Appendix I. Red List: Near Threatened. Population trend: Decreasing.

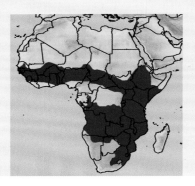

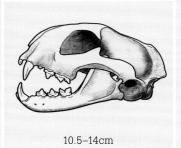

10.5–14cm

IUCN RED LIST (2015):
○ Least Concern (sub-Saharan Africa)
● Critically Endangered (North Africa)

Head-body length ♀ 63–82cm, ♂ 59–92cm
Tail 20–38cm
Weight ♀ 6–12.5kg, ♂ 7.9–18kg

Serval

Leptailurus serval (Schreber, 1776)

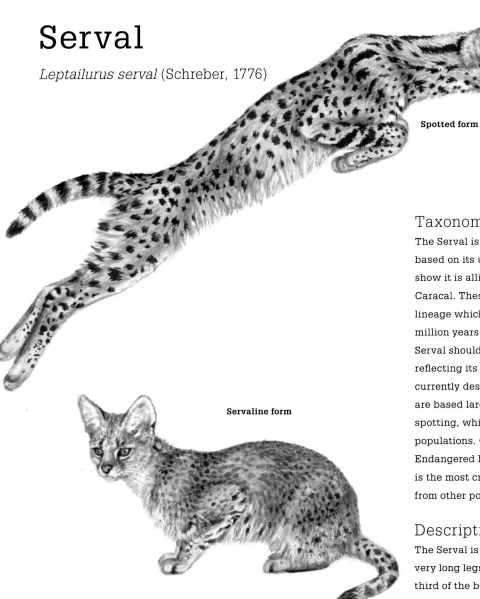

Spotted form

Servaline form

Taxonomy and phylogeny

The Serval is classified in its own genus *Leptailurus* based on its unique morphology. Molecular analyses show it is allied with the African Golden Cat and Caracal. These three species comprise the *Caracal* lineage which diverged as a distinct line around 8.5 million years ago. Some authorities argue that the Serval should be reclassified as *Caracal serval* reflecting its phylogeny. Seven subspecies are currently described but these are questionable; they are based largely on differences in colouration and spotting, which vary considerably within populations. Of putative subspecies, the Critically Endangered North African Serval *L. s. constantinus* is the most credible given long periods of isolation from other populations south of the Sahara.

Description

The Serval is a tall, medium-sized, slender cat with very long legs and a short tail measuring around a third of the body length. The smallish head is lightly built and dominated by very large, parabolic ears. Background colour is pale tawny to golden-yellow fading to pale underparts, and marked all over with bold, black spots that coalesce into long blotches on the nape, shoulders and limbs. A tawny-buff-coloured morph with faint, speckled spotting known as 'servaline' or 'Small-spotted Serval' was once considered a separate species but is merely a colour variant. Servaline individuals are known mainly from West and Central Africa especially at the savanna-rainforest ecotone, and there is one recent record (2013) from Kibale National Park, Uganda. Melanistic individuals are common in some populations, primarily in moist, highland areas of Ethiopia, Kenya

Relative to its size, the Serval has the largest ears of any felid. When resting relaxed, as in this adult female in Kenya, the ears almost meet in the middle of the head.

Right: **Although the Serval is rarely observed in trees, it is a capable climber when required. If dense ground cover is unavailable, Servals take refuge in trees when threatened by predators including people.**

Below: **Melanism in Servals occurs chiefly within five degrees of the Equator, with rare records from more distant populations. The reason for the pattern is unknown.**

and Tanzania, but also recorded occasionally from dry savanna woodland (Tsavo National Park and Amboseli region, Kenya) and rainforest-savanna mosaics (Batéké Plateau, south-eastern Gabon; Chinko Basin, south-eastern Central African Republic). All three forms – spotted, servaline and black – sometimes occur in the same population, for example in the Batéké Plateau and Chinko Basin.

Similar species The Serval is distinctive and not easily confused with other cats in the field; some servaline individuals have tawny, unspotted colouration that resembles the Caracal. Serval skins are sometimes mistaken for Cheetah.

Distribution and habitat

The Serval is endemic to Africa where it occurs widely throughout southern and East Africa, patchily in West Africa and as a relict population in North

Africa. It is naturally absent from most of the Congo Basin and the Sahara. Servals inhabit all types of savanna woodlands, grasslands and dry-humid forests, typically in close association with rivers, marshes, reedbeds and floodplains. They are naturally absent from dense, equatorial rainforest but they occupy forest-savanna mosaics and open patches in forested habitat. Servals do not inhabit true desert or semi-desert. They occur along watercourses in arid habitat, very rarely penetrating deep into dry areas; for example, a Serval was recorded in the Kalahari Desert, Kgalagadi Transfrontier Park, South Africa, in 1990. The relict population in coastal Morocco (and possibly Algeria) lives in arid shrub habitat interspersed with wetlands. Servals live up to 3,850m in alpine and subalpine moorlands, bamboo forests and grasslands in East Africa. They can live on farmland, where agriculture often gives rise to elevated rodent densities, but they avoid open monocultures devoid of cover. Servals are recorded from coffee, banana, sugarcane, eucalyptus and pine plantations.

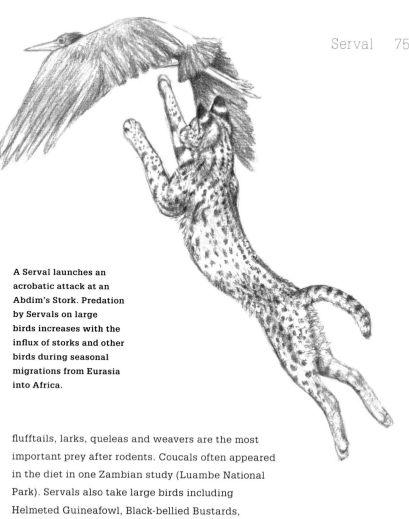

A Serval launches an acrobatic attack at an Abdim's Stork. Predation by Servals on large birds increases with the influx of storks and other birds during seasonal migrations from Eurasia into Africa.

Feeding ecology

Servals specialise in hunting small mammals in long-grass habitats. The Serval has proportionally the longest legs of any felid, 10–12cm higher at the shoulder than the similar-sized Caracal. This effectively raises the cat on stilts to move efficiently and quietly in long grass while it uses its huge, extremely sensitive ears to listen for hidden prey. Rodents and shrews comprise at least three-quarters of the diet in well-studied populations, peaking at 93.5 per cent of the diet for Servals on farmland in the KwaZulu-Natal Midlands, South Africa. The most commonly eaten species weigh 10–200g, and include vlei rats *Otomys* spp., multimammate mice *Mastomys* spp., grass mice, *Lemniscomys* spp., grooved-toothed swamp rats *Pelomys* spp., African Pygmy Mice and Nile Rats; larger rodents including ground squirrels, cane rats and Spring Hares, and lagomorphs (mainly Cape Hare and Scrub Hare), are occasionally killed. Small grassland birds like bishops, cisticolas,

flufftails, larks, queleas and weavers are the most important prey after rodents. Coucals often appeared in the diet in one Zambian study (Luambe National Park). Servals also take large birds including Helmeted Guineafowl, Black-bellied Bustards, herons, storks and flamingos. Reptiles and amphibians are opportunistically eaten, with frogs and toads comprising a high proportion of the diet in wetland habitats. Arthropods, especially large insects (locusts, grasshoppers and crickets), and freshwater crabs are readily eaten, though they generally contribute little to dietary intake. Rarely killed prey includes small carnivores such as Banded Mongoose and Large-spotted Genet; juvenile ungulates weighing up to 7kg, including Thomson's Gazelle, Oribi and duikers; and raptors, for example a Barn Owl and an unidentified falcon recorded from Zambia. A Serval in the Serengeti was observed killing a Side-striped Jackal pup before being chased off the dead pup by one of the parents. Livestock depredation is rare; they sometimes kill poultry and juvenile sheep or goats when unattended. Plant matter, including grass (probably as an emetic) and cultivated banana and avocado, is occasionally recorded in the diet.

Hunting is mostly nocturno-crepuscular and terrestrial; Servals rarely hunt in trees, usually only to pursue fleeing prey or rarely to probe nest-holes and crevices for reptiles and nestlings. Diurnal activity is more common during cold weather, for females with young kittens and in areas where Servals are not persecuted; they are largely nocturnal near people. Servals hunt chiefly by moving slowly and pausing frequently to listen; most prey is located by sound. Prey is captured by a high, arching pounce measuring up to 1.5m high and 3.6m long. They forcibly hit the target with forelegs swiftly extended, trapping and possibly stunning prey in one fluid motion. Servals take flushed birds and insects in flight; large birds are pulled from the air after a sprinting charge and high vertical leap that may exceed 2m. They sometimes 'fish' for rodents in their burrows and are reported to scratch at the plugged

entrances of mole-rat burrows to draw them to the surface. Most prey is killed with a precise bite to the head or nape. The Serval's skull is lightly built with relatively fine jaws reflecting its specialisation on small prey that is quickly killed. Dangerous snakes, including Puff Adder and Mozambique Spitting Cobra, are struck repeatedly with rapid blows to the head until stunned sufficiently for the Serval to safely bite the snake's head. Before eating, Servals pluck large birds and occasionally pull fur from larger mammal kills such as hares.

Servals in optimal, well-protected habitat in Ngorongoro Crater, Tanzania, make a kill in approximately 50 per cent of attempts; one female with kittens had a success rate of 62 per cent. Success does not differ between day and night hunts, though the relative proportion of prey types shifts depending on availability; for example, typically nocturnal species such as shrews are more commonly

The Serval's hunting technique has much in common with other rodent-specialists. The characteristic high-arching leap resembles the rodent-hunting pounce of many fox species, while the strike mirrors that of owls as they hit out explosively at the prey with clawed feet at the moment of impact.

killed at night. On average, a Serval makes 15–16 kills in a 24-hour period, making around 1.9–2.5 kills per kilometre covered. Based on intensive observations of habituated Servals in Ngorongoro Crater, an adult Serval kills an estimated 4,000 rodents, 260 small snakes and 130 birds in a year.

Servals seldom scavenge. On rare occasions, they have been observed confronting Black-backed Jackals and single Spotted Hyaenas over carcasses, and a Serval was seen to cautiously feed at night from a Lion-killed Zebra in Kenya as the Lions rested nearby.

Social and spatial behaviour

Very few Servals have been radio-collared, so spatial patterns are poorly understood. A four-year study in Ngorongoro Crater, Tanzania, of habituated Servals that were not collared has provided the most comprehensive information to date. Servals are normally solitary and territorial, though adults appear to be relatively tolerant of conspecifics – consorting pairs often travel and forage together, adult home ranges overlap considerably and aggressive confrontations are apparently rare. Adult males chase and attack young males although fatalities and serious injury have not been recorded. Servals (not collared, and therefore minimum estimates) in Ngorongoro Crater had ranges of 1.6–9.5km^2 (females) and 3.7–11.6km^2 (males). Range size of radio-collared Servals in KwaZulu-Natal, South Africa, was 15.8–19.8km^2 (two females) and 31.5km^2 (one male). Servals occur at relatively high densities. South African farmland, with intact, wetland habitat, has six to eight Servals per 100km^2 (Drakensberg Midlands). As many as 41 Servals per 100km^2 was estimated for optimum habitat in the Ngorongoro Crater during a prolonged period of high rainfall, but this was not based on robust methods.

At three months old, this Serval kitten is capable of making its own occasional kills of small prey but it is still entirely dependent on its mother for survival.

Reproduction and demography

Breeding appears to be weakly seasonal; Servals breed year-round but births tend to peak around wet season rodent eruptions in November–March (southern Africa) and August–November (Ngorongoro Crater). Gestation is 65–75 days; litters average two to three, exceptionally up to six (only recorded in captivity to date). Females are sexually mature at 15–16 months (captivity), males at 17–26 months (captivity). Females can breed until 14 (captivity); a wild breeding female of 11 is recorded. Kittens are independent at six to eight months old and often remain in the mother's range until 12–14 months.

Mortality Known predators include Leopard, Lion, Nile Crocodile and domestic dogs. Predation by Martial Eagles on kittens is recorded; a Serval female failed in an attempt to defend a kitten from a Martial Eagle in the Masai Mara. Predation on kittens by other large raptors, Black-backed Jackal, Spotted Hyaena and African Rock Python is likely, though there are no confirmed cases.

Lifespan Up to 11 years (female) in the wild, 22 in captivity.

Young Servals, such as these eight-week-old kittens, are vulnerable to a wide variety of predators. At this age, kittens are too young to accompany their mother while she hunts and she typically leaves them close to thick vegetation or other hiding places.

STATUS AND THREATS

The Serval is still widely distributed south of the Sahara and relatively common. It is extinct or relict in most of the north, west and extreme south of its original range. They are Critically Endangered in North Africa where very small populations survive in coastal Morocco and perhaps northern Algeria. Servals were extirpated from Tunisia but have been reintroduced into Feijda National Park using animals from East Africa. The species has expanded its range in a few regions; for example, they are gradually recolonising central South Africa in association with agricultural development and artificial water sources. Similarly, Servals possibly benefit from forest clearance and resulting encroachment of savanna habitats at the edges of the equatorial forest belt in Central Africa.

Habitat loss and associated persecution by people are the species' main threats. African wetlands and grassland savannas are under intense pressure for human use; draining, burning and overgrazing by livestock deplete both suitable habitat and rodent populations. Servals tolerate agricultural land provided there is sufficient cover and water, combined with tolerance from landowners. Servals are rarely responsible for significant conflict with livestock and poultry owners, and yet indiscriminate persecution means that the species is regularly killed on farmland. Servals are popular in the local fur trade and for fetish and traditional medicinal use, especially in north-eastern Africa, the West African Sahel belt and South Africa. Serval skin is popular among followers of the 'Shembe' Nazareth Baptist Church in South Africa for ceremonial *amambatha* (capes) in lieu of more expensive Leopard skin. The species is hunted for bushmeat in Republic of the Congo, Gabon and likely other Central and West African countries where rural populations depend heavily on bushmeat. The Serval is legally sport-hunted in all but 10 range countries (including Algeria where its presence is uncertain). CITES Appendix I. Red List: Least Concern (sub-Saharan Africa), Critically Endangered (North Africa). Population trend: Stable.

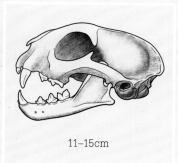

11–15cm

● **IUCN RED LIST (2008):** Least Concern

Head-body length ♀ 61–103cm,
♂ 62.1–108cm
Tail 18–34cm
Weight ♀ 6.2–15.9kg,
♂ 7.2–26.0kg

Caracal

Caracal caracal (Schreber, 1776)

Taxonomy and phylogeny

The Caracal was formerly classified in the genus
Lynx due to morphological similarities with the true
lynxes, but the resemblance is superficial and there
is no close relationship; the alternative common
names 'African Lynx' and 'Desert Lynx' are
misnomers. It is now classified with its closest
relative, the African Golden Cat, in the genus
Caracal. Together with the Serval, the genus
comprises the *Caracal* lineage that diverged around
8.5 million years ago. Nine subspecies of Caracal are
recognised, seven in Africa and two in the Middle
East and Asia, based largely on minor differences in
colouration which vary widely, and their validity
is dubious.

Description

The Caracal is a medium-sized, powerfully built cat
with a shortish tail that reaches the hind heels.
Muscular and slightly elongated hindlegs give the
hindquarters an elevated appearance, especially in
adult males. The head is heavily built with large,
distinctive ears, which are black-backed and liberally
flecked with white hairs, with very long, black tufts.
The tufts' function is unclear but they probably
enhance visual communication between Caracals

Right: **The function of the Caracal's extraordinary ear tufts is unclear. One theory suggests that the long tassels help to channel high-frequency sounds into the dish of the ear, although this has never been scientifically tested (C).**

Below: **A Caracal in mid-stalk on short-grass plains, Masai Mara National Reserve, Kenya.**

together with the bold, contrasting facial markings. Background colour varies from pale sandy-brown or pinkish-fawn to rich brick-red, with pale underparts. Caracals from arid regions tend to be pale, including very light sand-coloured individuals from the Middle East. The Caracal is largely unmarked, except for faint spots and blotches on the underparts in some individuals. Very dark chocolate-brown individuals occur rarely and true melanism is exceptional; three entirely black specimens are known from Kenya and Uganda. A dark grey morph, often with typically coloured extremities, occurs in Israel.

Similar species The Caracal is very distinctive. In Africa, it resembles the African Golden Cat at a glance, though the two species are largely divided by range. Caracals might be confused for uniformly coloured 'servaline' individuals of Serval in West and Central African savannas. Its lynx-like appearance might cause confusion in its Asiatic range, though it

co-occurs with the Eurasian Lynx only in a few scattered pockets along a very narrow band of sympatry across south-west Tajikistan, north-east Iran and south-east Turkey.

Distribution and habitat

The Caracal occurs in most of Africa except for true desert and rainforest regions; in southern Turkey and the Middle East excluding the Arabian Peninsula interior; and in south-west Asia from the east coast of the Caspian Sea to central India. Caracals have a broad habitat tolerance, able to occupy more open and more arid habitats than other, similarly sized cats. They favour all kinds of dry woodland savannas, dry forest, grasslands, coastal scrub, semi-desert and arid hilly or mountainous habitat. Caracals inhabit evergreen and montane forest in certain regions, typically in forested pockets surrounded by more open habitat, for example in the

Ethiopian Highlands. They do not inhabit open, true desert areas of the Arabian Peninsula, Namib and Sahara, but they occur in rocky habitat, inselbergs and watercourses located deep into very arid habitat. They do not occur in equatorial forests of Central and West Africa and are largely absent from a wide transitional band of forest-savanna mosaic surrounding the Congo Basin. Caracals are recorded to 2,260m in the Atlas Mountains, Morocco, and exceptionally to 3,300m in the Ethiopian Highlands. They tolerate pastoral and agricultural landscapes provided there is cover, and are recorded from eucalpytus and pine plantations.

Feeding ecology

The Caracal is a formidable hunter with very muscular, elongated hindlegs enabling explosive bursts of speed and spectacular, vertical leaps. Compared to the similarly sized Serval, the forepaws

Among similar, mid-sized felids, the Caracal is found in more open and arid habitats than any other species. This individual was photographed at the edge of Etosha Pan in Namibia where spring-fed grass and sedges provide suitable cover.

are relatively massive with well-developed claws, and the skull, dentition and biting temporalis and masseter muscles are very robust. These adaptations reflect the Caracal's prodigious ability to bring down large mammals up to four times its weight, though few populations specialise in oversized prey. Generally, frequent kills of small mammals weighing less than 5kg are supplemented by occasional, larger kills weighing up to 50kg. Mammals comprise 69.8 per cent (West Coast National Park, South Africa) to 93–95 per cent (Mountain Zebra National Park, South Africa) of the diet. In South Africa's Cape provinces, the most important prey items are Rock Hyrax, rodents, Klipspringer, Grey Rhebok, Cape Grysbok, Steenbok, Mountain Reedbuck, Springbok and duikers. Almost 60 per cent of the diet in the Kgalagadi Transfrontier Park (South Africa) is made up of Spring Hare, Highveld Gerbil, Brants's Whistling Rat and Striped Mouse. The most important prey in Turkmenistan is Tolai Hare, Great

Gerbil and various jerboa species. The largest recorded prey of Caracals includes adult Bushbuck, Impala ewes and young Greater Kudu. Juvenile and yearling Barbary Sheep, Dorcas Gazelle and Goitered Gazelle are recorded from the Sahara and Asia. Small carnivores are opportunistically preyed upon, the largest confirmed species being Black-backed Jackals and Red Foxes; 10 per cent of the diet in the Kgalagadi Transfrontier Park comprises Bat-eared Fox, Cape Fox, Yellow Mongoose, African Wild Cat and Striped Polecat. Domestic cats and Egyptian Mongooses are recorded as occasional prey in Israel.

After mammals, the most important prey category is made up of birds, mainly guineafowls, francolins, quails, partridges, sandgrouse, doves, pigeons and small passerines. Larger species, including Kori Bustard and Indian Peafowl ('peacock'), are also killed and even the Ostrich is recorded in Caracal scats. It is unclear how often they kill adult Ostriches, but there is one reliable record of an adult taken while bedded down. Large raptors, including Martial Eagle, Steppe Eagle and Tawny Eagle, are killed rarely, predominantly while roosting at night. Reptiles up to the size of large snakes and monitors are eaten; reptiles comprise 12–17 per cent of the diet in West Coast National Park. Amphibians, fish and invertebrates are occasionally consumed. Beetles appear in around a quarter of all Kgalagadi Caracal scats but they contribute a negligible amount to actual intake. Caracals kill small livestock, most often where stock is free-ranging and unattended, combined with wild prey being scarce; livestock comprises 3.6–55 per cent of the diet in farming areas in southern Africa. Domestic fowl are readily taken.

Caracals generally hunt at night and crepuscularly. Diurnal activity is more common during cooler periods, often reflecting reduced prey densities during the lean season and/or diurnal activity of preferred prey; for example, Kgalagadi Caracals rely considerably on diurnal Brants's Whistling Rats. Caracals are also more likely to be diurnal when protected from persecution. Most

Caracal populations on the Arabian Peninsula are restricted to coastal mountainous and desert habitat where their long-term future is in doubt. This individual was photographed in the Hawf Protected Area, Yemen.

hunting takes place on the ground but the Caracal is a very capable climber, and the kills of roosting birds high in trees testify to some arboreal hunting ability. Caracals locate prey primarily by sight and hearing, followed by stalking to within 5m or waiting in ambush before rushing the prey. Based on reconstructing hunting sequences from tracks, two-thirds of hunts in the Kgalagadi involve no stalk at all; on sighting prey, these Caracals hunt primarily by waiting in ambush before initiating a chase as prey comes close, or launching directly into a chase as soon as the prey is spotted. The Caracal's final sprint is explosive – the species is often regarded as the fastest of smaller cats, though there are no reliable measurements. Caracals chase prey for up to 379m but most chases are considerably shorter, averaging 12m for small prey and 56m for large prey (Kgalagadi). The Caracal is marvellously athletic, able to leap at least 2m high from a stationary position and at least 4.5m long while running. Its

ability to catch birds in flight is well known, occasionally knocking down several birds in one spectacular leap, a skill for which it was formerly tamed by various Asian cultures. Small prey species are killed by biting the skull or nape while ungulates are asphyxiated with a throat bite. This is often accompanied by powerful raking of the prey's belly or chest with the claws of the hindfeet, leaving distinctive, deep claw marks on carcasses. A combination of two bites, to the nape and throat, has been observed, for example by Kgalagadi Caracals when killing African Wild Cats, perhaps indicating the Caracal shifting position depending on the defence mounted by the victim.

Hunting success rates are poorly known. Kgalagadi Caracals kill prey on 10 per cent of hunts based on spoor-tracking, but this method probably underestimates kill rates of small prey when no remains are left. The success rate for large prey in the same study was 20 per cent. On average, these

A Caracal appropriates the carcass of a dead Bat-eared Fox from two Black-backed Jackals despite a determined defence mounted by the jackals, Kgalagadi Transfrontier Reserve, South Africa.

Caracals make a hunting attempt every 1.6km and make a kill every 16.3km. In Sariska Tiger Reserve, India, each Caracal is estimated to eat 2,920 to 3,285 rodents per year.

Caracals cache large kills in thick cover and return to feed for up to four or five nights. Larger prey are seldom dragged far but are eaten on site or moved a few metres if cover is available. Caracals do not eat the main bones of large carcasses, often leaving the entire, articulated skeleton of ungulates with all soft tissue and small or cartilaginous bones consumed. Caracals readily scavenge, including by kleptoparasitism, especially from Black-backed

Jackals (and vice versa). Camel and gazelle carcasses in Saudi Arabia are an important food source, and Cape Fur Seals are scavenged in coastal southern Africa. Hoisting kills into trees has been observed very rarely, perhaps in response to an immediate threat from another carnivore.

Social and spatial behaviour
Caracal ecology is poorly studied; limited information from radio-collaring studies in Israel and southern Africa shows a typical felid pattern. Adults are largely solitary and maintain enduring home ranges, usually with exclusive core areas and

At around six months old, these two Caracal kittens will remain with their mother (on the far left) for another three to six months before they disperse. Masai Mara National Reserve, Kenya.

considerable overlap at the edges. Caracals display behaviours indicative of territorial defence such as scent-marking and scraping. Adults, especially males, often have extensive scarring to the face and ears presumably from intraspecific fighting.

Territories are large, though the telemetry data is biased towards drier habitats where larger ranges are to be expected. The smallest recorded ranges occur in the relatively mesic, coastal areas of the Western Cape, South Africa, ranging between 3.9 and 26.7km² (females) and 5.1–65km² (males). In the Arava Valley, Israel, ranges average 57km² (females) and 220km² (males). Based on very limited locations, Kgalagadi Caracals have average ranges of 67km² (females) to 308km² (males), while three Central Namibian males used 211.5–440km². An adult Saudi male collared for 11 months used 865km² with no evidence of a preferred core area (Harrat al-Harrah Protected Area, Saudi Arabia). There are few reliable density estimates – in a small protected area with abundant rodents and no competing large carnivores, adult Caracals reach approximately 15 individuals per 100km² (Postberg Nature Reserve).

Reproduction and demography

Caracals reproduce year-round, with weak seasonality in climatically moderate parts of their range – births peak in October–February (South Africa) and November–May (East Africa). Seasonality may be more pronounced in areas with greater seasonal extremes. Oestrus lasts one to six days during which the male and female remain together. Up to three males have been recorded attending a receptive female, each of which was mated. Gestation is 68–81 days. Litters average two to three kittens, exceptionally reaching six. Kittens reach independence at 9–10 months old, and seem to follow a typical felid pattern of female philopatry and male dispersal. A male in Israel dispersed 60km and a young Kgalagadi male moved more than 100km in five months before being shot. Sexual maturity is 12–16 months for both sexes, and a captive female reproduced at 18 years.

Mortality Adult Caracals are occasionally killed by larger carnivores, mainly Lions and Leopards. Black-backed Jackals are recorded preying on young kittens, which are presumably vulnerable to a wide array of additional predators. Infanticide by males occurs, though the circumstances are unknown.
Lifespan Unknown in the wild, 19 years in captivity.

In exposed habitats, Caracals are vulnerable to kleptoparasitism from scavengers. Although they defend their kills aggressively, they are usually unsuccessful against the much larger Spotted Hyaena.

STATUS AND THREATS

The Caracal is widely distributed and relatively common in southern and East Africa. It is rare or relict in Central, West and North Africa but may be relatively secure in large protected areas. It is relatively widespread throughout the Middle East and Asian parts of its range but it probably always occurred there in low densities, now exacerbated by human threats so that the species is considered endangered in much of Asia.

Habitat degradation, loss of prey and persecution by people are significant threats in Central, West and North Africa and most of Asia. Caracals are hunted intensively on livestock land, especially in Namibia and South Africa where the Caracal is typically classified as a 'problem animal' outside protected areas and killed without restriction. Despite this, they are resilient in southern Africa, probably because of the high habitat suitability and prey availability combined with the lack of larger carnivores, and the Caracal is difficult to extirpate there. The Caracal's ability to move large distances enables it to rapidly fill spaces created by chronic persecution. They are sport-hunted with few restrictions in much of East and southern Africa.
CITES Appendix I (Asia), Appendix II (Africa). Red List: Near Threatened. Population trend: Least Concern.

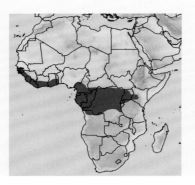

12.6–14.6cm

IUCN RED LIST (2015):
Vulnerable

Head-body length 61.6–101cm
Tail 16.3–37cm
Weight ♀ 5.3–8.2kg, ♂ 8–16.0kg

African Golden Cat

Caracal aurata (Temminck, 1827)

Taxonomy and phylogeny

The African Golden Cat was formerly grouped with
the Asiatic Golden Cat due to superficial physical
similarities, but molecular analyses show they are
not closely related. Its closest relative is the Caracal
with a divergence estimated around 1.9 million years
ago, and these two species are classified together in
the genus *Caracal*. It is somewhat more distantly
related to the Serval. These three species
comprise the *Caracal* lineage, which diverged
around 8.5 million years ago. Two subspecies
of African Golden Cat are currently described,

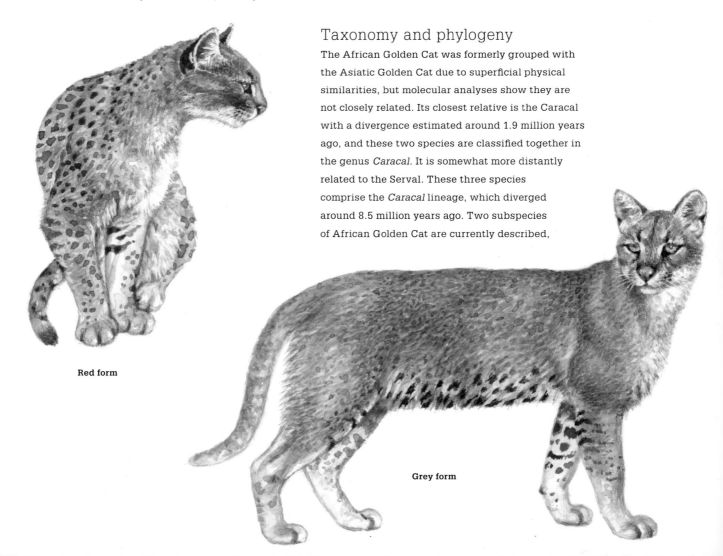

Red form

Grey form

representing populations west of the Cross River, Nigeria, and east of the Congo River, Democratic Republic of the Congo. These are based largely on superficial differences in pelage and require molecular analyses for confirmation.

Description

The African Golden Cat is a medium-sized, solidly built cat with a medium length tail that reaches just below the hind heel. The head is relatively robust with a short face, heavy muzzle and rounded ears with greyish-black backs. Kittens are sometimes born with small tufts on the ears, likely a sign of the species' shared ancestry with the Caracal; the tufts are lost after a few months and are not seen in adults. Background fur colour is highly variable but clusters into two distinct colour forms, reddish-brown and grey; both are highly variable ranging from sandy-brown to mahogany-brown, and silver-grey to blue-grey, with intergradation between them, giving rise to a confusing number of described forms. Both colours can be born into the same litter.

Left: **The African Golden Cat has a distinctive, heavy-muzzled profile that resembles a small leopard. Many local names reflect the similarity, translating as 'son of the leopard' and 'the leopard's brother' (C).**

Below: **In much of its range, the African Golden Cat co-occurs with the Leopard which is the only large cat resident in African rainforest. Where Leopards have disappeared, the Golden Cat has become the top carnivore such as in Kibale National Park, Uganda where this individual was photographed.**

Above: **An African
Golden Cat attacks a
Red Colobus Monkey.
Recent video-trap
footage of such an
attack confirms
that Golden Cats
attempt to take prey
that significantly
outweighs them.**

Spotting is also highly variable, ranging from large, rosette-like blotches to faint freckling or an absence of markings altogether except on the underparts. Individuals in the west of the range (e.g. in Gabon) tend to show greater variation in spotting with extensively spotted individuals being more common. Eastern Golden Cats (e.g. in Uganda) tend to be more uniformly, lightly spotted. Melanistic individuals have been recorded from Democratic Republic of the Congo, Liberia and Uganda at low frequencies (fewer than 10 per cent of the population) and may be most common in high-altitude sites. Individual Golden Cats do not change colour, a misconception stemming from a questionable report of an ailing, captive animal that apparently changed from reddish to grey and died soon after. Darkening to a deeper shade of the same colour has been observed in captive individuals as they age.

Opposite: **A red
African Golden Cat,
photographed in a
logging concession near
Ivindo National Park,
Gabon, showing an
intermediate amount
of spotting. Camera-
traps (such as the unit
in the background)
are placed along trails
used by Golden Cats to
secure images without
disturbing the cat.**

Similar species Superficially similar to the Asiatic Golden Cat, but they are not sympatric in the wild. African Golden Cats are somewhat similar to Caracals, but this species is absent from rainforest and therefore most Golden Cat range. The two species are sympatric in less than 6 per cent of the Golden Cat's range at its very limits, for example in southern Central African Republic.

Distribution and habitat

The Golden Cat is endemic to equatorial Africa where it occurs in two disjunct populations in West and Central Africa, separated by the dry Dahomey Gap. They occur east of the Western (Albertine) Rift Valley but reports east of the Eastern (Gregory) Rift Valley are unconfirmed. African Golden Cats are strongly associated with moist forests including bamboo, montane and subalpine forest, and forest mosaics from the coast to 3,600m. They inhabit riverine forest strips in savanna woodlands, allowing them to penetrate drier, more open habitat they would otherwise avoid. They also occur in dense wooded savanna and savanna-forest mosaics; they are occasionally seen crossing open savannas in mosaics. They are tolerant of some human-modified habitats, including banana plantations in forest and abandoned, recently logged areas with secondary undergrowth.

Feeding ecology

This is the most poorly known cat species in Africa, and the African Golden Cat's diet is known largely from analysing scats. The most important prey species are thought to be small mammals up to 5kg, primarily various mice, rats, squirrels and African Brush-tailed Porcupines as well as insectivores such as shrews and sengis. Small forest antelopes,

particularly Blue Duikers, are also killed and may contribute more to the diet by weight in some areas. Golden Cats also hunt tree pangolins, small primates such as galagos and larger primates to the size of guenons (4–6kg) and colobus monkeys (4–6kg). One of the few eyewitness accounts of a hunting Golden Cat reported a Sykes' Monkey killed on the ground. In two separate camera-trap records from 2014 in Kibale National Park, Uganda, a Golden Cat carrying a freshly killed, young adult Red-tailed Monkey was photographed and a small Golden Cat, possibly an adult female, was filmed in a spectacular attack on Red Colobus Monkeys feeding on the ground; the cat caught an adult monkey much larger than itself but the monkey escaped. Bushpigs have been recorded in the diet but it is unknown if they were killed (only juveniles would be vulnerable) or scavenged. Second to small mammals, birds are the most important component of the diet, particularly large ground-

Left: **African Golden Cats occupy dense habitat in which low light is the norm, even during the day. Nocturnal visual acuity is probably greatest among forest-dwelling felids compared to those species inhabiting mainly open habitat.**

Like many felids, African Golden Cats readily travel along roads and trails for hunting opportunities as well as to scent-mark prominent trees and bushes.

socio-spatial system, in which male ranges encompass all or part of the ranges of several adult females. Adults prefer to move along roads and well-defined trails, on which they urine-mark and leave faeces in prominent places, which are hallmarks of territorial behaviour. There are no reliable estimates of range size. The only rigorous density estimates come from central Gabon, ranging from 3.8 cats/100km^2 in unprotected, mature forest with moderate levels of bushmeat hunting by people, to 16.2 cats/100km^2 in well-protected, undisturbed forest (Ivindo National Park).

Reproduction and demography

Very poorly known in the wild. A single record of gestation from captivity was 75 days. Three litters born in captivity all numbered two kittens, which were weaned around six weeks. Sexual maturity in captive animals is 11 months (females) and 18 months (males).

Mortality Poorly known; Leopards are confirmed predators.

Lifespan Unknown in the wild, up to 12 years in captivity.

- -

STATUS AND THREATS

Golden Cats are thought to be naturally rare, though recent camera-trapping suggests moderate to high densities in suitable habitat, including secondary and selectively logged forests. They are more tolerant of forest modification than once thought but they are nonetheless dependent on forest cover. Conversion of forest to open habitat in concert with depletion of natural prey are the most serious threats in much of the range, especially at the extremities. Bushmeat hunting in West and Central Africa heavily impacts prey species and Golden Cats themselves are killed frequently in some areas, as bycatch or intentionally for bushmeat and fetish markets.

Cites Appendix II. Red List: Vulnerable. Population trend: Decreasing.

- -

dwelling species such as francolins and guineafowl. Golden Cats doubtless eat various herptiles, though there are very few records; reptile scales occurred in 11 of 205 scats from Gabon. They presumably take unguarded poultry, though rural people interviewed in Gabon and Kibale, Uganda, do not consider them a nuisance. They are occasionally blamed for the loss of small goats at the edge of Kibale National Park (where they are now the largest carnivore except for domestic dogs), though unequivocal evidence is lacking. Hunting is mainly terrestrial and camera-trap surveys show they are active at all times of day and night, with peaks at dawn, midday and late afternoon to dusk. Golden Cats occasionally scavenge from dead animals in wire snares and are thought to scavenge eagle kills on the forest floor.

Social and spatial behaviour

Camera-trap images are mostly of solitary adults and it is assumed Golden cats have a typical felid

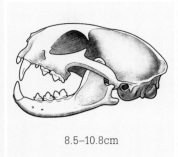

8.5–10.8cm

IUCN RED LIST (2015): Least Concern

Head-body length ♀ 43–74cm,
♂ 44–88cm
Tail 23–40cm
Weight ♀ 2.6–4.9kg, ♂ 3.2–7.8kg

Geoffroy's Cat

Leopardus geoffroyi (d'Orbigny & Gervais, 1844)

Taxonomy and phylogeny

Geoffroy's Cat is classified in the *Leopardus* genus and is most closely related to the Guiña with a shared ancestor estimated under a million years ago. Both species were formerly classified in the genus *Oncifelis* with the Colocolo but they are now firmly considered to represent a closely related sub-branch of the *Leopardus* genus together with the oncillas. Geoffroy's Cats hybridise with the recently described Southern Oncilla (*L. guttulus*) in southern Brazil where the two species' ranges overlap; this is an active hybridisation zone meaning that hybrids are fertile and interbreeding is ongoing, reflecting a very close evolutionary relationship between the two species. There are up to four subspecies described for Geoffroy's Cats, but preliminary genetic analysis suggests few molecular differences across the range.

Description

Geoffroy's Cat is the largest of the South American temperate zone small felids, reaching the size of a large domestic cat (Andean Cats may be comparable based on very few weighed individuals). Males are 1.2–1.8 times the weight of females. Body size varies considerably across the range (but not in a strong cline increasing from north to south, as often reported) possibly related to prey availability. Based on relatively few captured animals, average weight varies from 3.1kg (females)–3.7kg (males; Uruguay), 4.1kg (females)–5kg (males; Parque Nacional Torres del Paine, Chile), to 4.2kg (females)–7.4kg (males; Campos del Tuyú Wildlife Reserve, Argentina). The fur colour is variable, ranging from rich yellow-brown to pale buff and silver-grey; southern animals are typically pale coloured while richer tawny or reddish tones are more common in the north of the range. The body is covered with small, solid dab-like dark brown or black spots coalescing to blotches on the nape, chest and lower limbs. The tail has 8–12 narrow bands interspersed with small spots; the bands become wider towards the tail tip, which is dark.

Melanism is common, particularly in Uruguay, south-eastern Brazil and eastern Argentina, but is generally rare elsewhere in the range.

Similar species Geoffroy's Cat is very similar in appearance to the closely related Guiña, which is considerably smaller, generally has richer colouration and a distinctive bushy tail. The two species are known to overlap only at the extreme eastern edge of the Guiña's range, for example Parque Nacional Los Alerces, southern Argentina, and Parque Nacional Puyehue, Chile.

Distribution and habitat

Geoffroy's Cats occur from central Bolivia through western Paraguay, extreme south-eastern Brazil and Uruguay and in most of Argentina to the Strait of Magellan in Chile; their distribution stops at the eastern Andes and extends marginally into southern Chile along the border with Argentina. Geoffroy's Cats inhabit a wide variety of habitats including all kinds of subtropical and temperate brushland, woodlands, dry forest, semi-arid scrub, *pampas* grasslands, marshlands and alpine saline deserts, from sea level to 3,300m in the Andes. They occur in open grassland but typically in areas with wooded and brushy patches or well-vegetated marshland. They are not found in tropical or temperate rainforest. They occur in modified landscapes with cover such as ranchlands with scrub-grassland mosaics. They also occur in conifer plantations especially those with remnants of native vegetation – Geoffroy's Cats collared in Ernesto Tornquist Provincial Park, central Argentina, avoided natural grasslands which were degraded by the presence of feral horses, and mainly used overgrown exotic woodlands outside the park. They are recorded using abandoned houses as shelters in *pampas* croplands.

Feeding ecology

Geoffroy's Cat is a versatile generalist in which small vertebrates make up 78–99 per cent of the diet, the species composition of which varies according to the region and prey availability. In most populations, the diet is dominated by small rodents weighing up to 200–250g, including grass mice, rice rats, marsh rats and cavies, and small passerine birds; consumption of the latter typically increases during spring-summer migrations. Larger prey dominates according to the season and the site. Introduced

European Hares (~2.5–3.2kg) have become important prey in many areas. Hares comprise more than half of the diet in southern Chile where Geoffroy's Cats are large, but less so where cats are smaller; for example, only 2 per cent of the diet in Parque Nacional Lihue Calel, Argentina. Large waterbirds with an average weight of 1.3kg are most important for cats in coastal lagoon habitat (Mar Chiquita, Argentina) where 12 species are eaten, including Neotropic Cormorants, White-faced Ibises, coots, ducks, as well as occasional kills of larger species including Chilean Flamingos and Coscoroba Swans. Waterbirds are most important during spring when their abundance peaks; the diet shifts more to small rodents and hares in summer and autumn when birds migrate and their abundance declines. Incidental prey across the range includes Coypus (introduced), Six-banded Armadillos, hairy armadillos, tree porcupines, small opossums, small reptiles, amphibians, crabs, fish and invertebrates (mainly beetles, which contribute little to energetic requirements). They readily raid domestic poultry; sheep recorded in the diet (for example, in southern Brazil) is probably scavenged.

Geoffroy's Cats are primarily nocturno-crepuscular throughout their range, with activity increasing as dusk approaches and peaking between 21:00 and 04:00. During a severe drought and collapse of prey populations in and around Lihue Calel in 2003, radio-collared Geoffroy's Cats were primarily diurnal, presumably in search of food. They returned to nocturno-crepuscular foraging patterns when the drought had passed, with 93 per cent of activity records occurring from 20:00 to 06:00. Geoffroy's Cats mostly forage on the ground in search of small rodents and birds in ground vegetation; although they climb very well, there are no observations of arboreal hunting. They readily swim and hunt in marshy habitat; Coypus, marsh rats, birds, frogs and fish are taken at the water's edge. Geoffroy's Cats in coastal lagoon habitat launch attacks on waterbird roosting areas from dense grassy vegetation in the shallows at the edges

A Geoffroy's Cat killing a wild cavy with a nape bite, an efficient technique used by all cats to rapidly dispatch small prey. Large prey, with proportionally broader, more muscular necks, are less vulnerable to a nape bite and are usually killed by a throttling bite to the throat.

of colonies. Geoffroy's Cats cache large kills. They were observed twice hauling European Hare carcasses into *Nothofagus* trees in Chile, and a female Geoffroy's Cat in Argentina failed in an attempt to haul a Red-legged Seriema 4m into a tree; she later stored the carcass in a burrow which she also occupied.

Social and spatial behaviour

Geoffroy's Cat is solitary; males occupy larger home ranges than females, and one male range typically overlaps multiple female ranges. Some radio-tracking data show that home ranges are not stable and that Geoffroy's Cats readily abandon their ranges to become transient, but it is likely that those populations were monitored during periods of extreme ecological stress and high population turnover, for example, in Parque Nacional Lihue Calel where only 11 per cent of the cats identified during a

protracted drought in 2006 were found two years later. In Torres del Paine National Park, Chile, an adult female maintained the same range for three years and a young female captured as a juvenile was still in the same area two years later. The degree of territorial defence is unknown but Geoffroy's Cats mark their ranges assiduously; unusually for felids, they often deposit faeces in certain trees which they repeatedly mark over time, creating conspicuous, arboreal middens. For example, 93 per cent of 325 faeces collected in Torres del Paine were in arboreal middens 3–5m high, typically where the main trunk diverged into several branches creating a natural platform or bowl. Middens on the ground are also used. Across five protected areas in Argentina, 47.6 per cent of all defecation sites were in trees: 38.1 per cent were on the ground, mainly along trails, and the remainder were in burrows.

Range sizes average 1.5–5.1km² (females) and

Although not nearly as richly marked as other neotropical cats such as the Margay and Ocelot, Geoffroy's Cat was formerly killed in huge numbers for its spotted fur. It is now fully protected across its range although some illegal local hunting for fur still takes place (C).

Geoffroy's Cat is very
unusual among felids
in depositing faeces in
latrines called middens;
it is even rarer that most
latrines are in trees such
as the *Nothofagus* shown
here. Latrine use is also
known from the related
Colocolo and Ocelot
although it is less common
and largely terrestrial.

December and May. In captivity, they breed year-round and there is no evidence for seasonality in the northern part of the range. In captivity, gestation lasts 62–78 days and litter size is one to three kittens. Wild females den in burrows (probably created usually by armadillos), in thick vegetation and possibly in tree cavities (which are definitely used as rest sites by adults). Kittens achieve adult size (but not weight) at around six months of age but sexual maturity is surprisingly late for a small cat, at 16–18 months (captivity).

Mortality Pumas are known predators and there is one case of probable predation by a Culpeo Fox. Severe droughts with collapses in prey numbers have been shown to elevate mortality considerably due to starvation combined with high parasite loads. Where studied, people and domestic dogs are usually the main source of mortality.

Lifespan Unknown in the wild, up to 14 years in captivity.

2.2–9.2km² (males); the largest ranges are found in disturbed habitat (agricultural-woodland mosaic, east-central Argentina). Per day, radio-collared cats move on average 583m (females; up to 1,774m) to 798m (males; up to 1,942m) in well-protected Argentine *pampas*, compared to 680m (females; up to 2,859m) to 1,213m (males; up to 3,704m) in disturbed agricultural-woodland mosaic. Rivers do not restrict their movements; a radio-collared female in Chile regularly crossed a 30m wide fast-flowing river in her territory, and two young males swam the same river when dispersing. Density estimates range from four cats per 100km² (Argentine *pampas* during an extreme drought and prey shortage) to 45–58 cats per 100km² in ranchlands of the Argentine Espinal. A very high estimate of 139 cats per 100km² from protected Argentine scrublands may be an overestimate.

Reproduction and demography

Poorly known from the wild, but Geoffroy's Cats are considered to breed seasonally in the southern part of their range where winters are very cold; based on limited observations, births occur between

STATUS AND THREATS

Geoffroy's Cats are widely distributed and are found in a broad range of habitats including some highly disturbed landscapes. They reach high densities in good habitat and are considered common in much of their range. However, threats are poorly understood. They were killed in huge numbers for their furs throughout the 1970s and 1980s, particularly in Argentina which exported at least 350,000 skins from 1976 to 1980. Their abundance was a factor in the massive trade, but there is no doubt the trade was unsustainable and led to reductions and extirpation in some areas. International trade in spotted cat furs ceased in the late 1980s and the fur trade is no longer a major threat, although domestic (and illegal) use of their skins continues in some places. The primary threat today is habitat conversion, for example to agricultural monocultures that even Geoffroy's Cats cannot occupy. Additionally, they are often killed on roads, by domestic dogs, and by people for depredation on poultry, all of which probably exacerbates the threats to populations already under pressure in modified landscapes. Geoffroy's Cats in two Argentinean protected areas (Parque Nacional Campos del Tuyú and Parque Nacional Lihue Calel) tested positive for exposure to a wide variety of domestic carnivore diseases and parasites, though no effects on individuals or populations were observed.

CITES Appendix I. Red List: Least Concern. Population trend: Decreasing.

Northern Oncilla

Southern Oncilla

● **IUCN RED LIST (2008):**
Vulnerable

Head-body length ♀ 43–51.4cm,
♂ 38–59.1cm
Tail 20.4–42cm
Weight ♀ 1.5–3.2kg, ♂ 1.8–3.5kg

Northern Oncilla

Leopardus tigrinus

(Trigo, Schneider, de Oliveira, Lehugeur, Silveira, Freitas & Eizirik, 2013)

Southern Oncilla

Leopardus guttulus

(Schreber, 1775)

Little-spotted Cat, Tiger Cat

Taxonomy and phylogeny

Oncillas belong in the *Leopardus* genus, and they have intriguing evolutionary relationships with the other cats of this lineage. Until 2013, the Oncilla was considered a single species but recent genetic analysis has revealed two closely related 'cryptic'

6.8–8.5cm

species that have been evolutionarily separate for at least 100,000 years. Oncillas in north-eastern Brazil (*L. tigrinus*) are distinct from those in southern Brazil (now considered *L. guttulus*), which presumably also applies to populations in the Amazon Basin and northern South America. Their respective ranges meet in central Brazil where the two species overlap but apparently do not interbreed. Southern Oncillas hybridise with Geoffroy's Cats in the wild while Northern Oncillas do not, and yet they show evidence of historical hybridisation with the Colocolo. Oncillas elsewhere in Latin America were not included in the analyses, but earlier sampling showed that the Costa Rican population was distinct from southern Brazilian oncillas, and it is currently considered a discrete subspecies, the Central American Oncilla *L. tigrinus oncilla*. It is possible that further research will describe more cryptic species.

Description

The oncillas are the second smallest tropical Latin American cats, with a slender, lightly built body about the size and proportions of a young, lean domestic cat. The fur colour ranges through shades of pale to dark buff or ochre marked with orderly rows of black or dark brown blotches or small rosettes with a coffee-brown or reddish centre. The two recently described species are extremely similar

Outwardly, the two oncilla species are extremely difficult to distinguish, especially above the neck. Both have a small, very lightly built head suited for very small prey that rarely weighs more than 500 grams (C).

in appearance. There is apparently a slight trend for *L. tigrinus* to be more lightly coloured with smaller rosettes than *L. guttulus*. Melanistic individuals are recorded.

Similar species Oncillas are very easily confused with the Margay, which is larger, more richly marked and has a distinctively longer tail.

Distribution and habitat

Oncillas occur from northern Venezuela to southern Bolivia, eastern Paraguay, far southern Brazil and extreme northern Argentina; the boundary between the two newly described species appears to be central Brazil (and possibly across the Bolivian range, which is likely to be *L. guttulus*). There is a disjunct population in the central cordillera of Costa Rica and north-west Panama (the Central American Oncilla, which might comprise a third species). Both Oncilla species occur in a broad range of habitats including all types of forest, woodlands, wet and dry savannas, arid scrublands and coastal *restinga* (scrub on sandy beaches). They live from sea level to 3,200m, and exceptionally at 3,626m (Costa Rica) to 4,800m (Colombia). In Central America, they are restricted to oak-dominated cloud and elfin forests above 1,000m. In Brazil, there is speculation that Northern Oncillas *L. tigrinus* occur mainly in open, dry habitats while Southern Oncillas *L. guttulus* are mainly forest-living. They are strangely absent from or very rare in Amazon Basin lowland rainforest, though this might reflect poor sampling. Oncillas can occupy degraded habitats close to people including rangelands, plantations, agricultural mosaics and peri-urban areas even near large cities, provided there is dense cover.

Feeding ecology

Oncilla prey is typically very small, averaging 100–400g and is not recorded exceeding 1kg. Typical prey includes small rodents, shrews, small opossums,

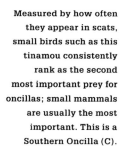

Measured by how often they appear in scats, small birds such as this tinamou consistently rank as the second most important prey for oncillas; small mammals are usually the most important. This is a Southern Oncilla (C).

small birds and reptiles. Invertebrates often appear in scats but contribute very little to energetic requirements. Mid-sized diurnal lizards (ameivas and small iguanas) dominate the diet in semi-arid *caatinga*, north-eastern Brazil, where rodent densities are low. There are few direct observations of hunting but the species' prey profile indicates that they forage mainly terrestrially and nocturno-crepuscularly but with some flexibility depending on prey activity (and possibly the presence of Ocelots which they might avoid). Oncillas are more diurnal in the *caatinga* reflecting their reliance on diurnal reptiles. They rarely take poultry.

Social and spatial behaviour

Oncillas likely follow a typical felid solitary socio-spatial pattern but their ecology is poorly known. Range size has been calculated for eight individuals only from Brazilian savannas and agricultural-forest mosaics where estimates are 0.9–25km² (females)

and 4.8–17.1km² (males). Density estimates from camera-trapping suggest they are naturally rare compared to other small felids in Latin America: 0.01 per 100km² (lowland Amazon forest) to 1–5 per 100km² in other areas of the range. They apparently reach 15–25 per 100km² in areas where Ocelots are absent, presumably due to release from competition and predation.

Reproduction and demography

Unknown from the wild. Limited information from captive individuals indicates gestation is 62–76 days and litter size is one and rarely two kittens.

Mortality Poorly known – there is a single record of an Ocelot killing an adult, and a Brazilian female died of heartworm disease. Domestic dogs are likely to be a significant predator in anthropogenic landscapes.

Lifespan A female lived to 17 years in captivity, but longevity is certain to be much less in the wild.

A pair of Northern Oncillas during courtship, the female signalling her receptivity by lordosis in which she lowers her forequarters and raises the hips. There is no recorded observation of wild oncillas mating but the behaviour is assumed to follow a typical felid pattern.

STATUS AND THREATS

Oncillas are camera-trapped at very low rates in most areas and are regarded as naturally rare. Habitat loss is the main threat. Although some populations occupy modified habitats, they are closely associated with dense cover especially in Central America and the Andean forests of Colombia and Venezuela where oncillas have become locally extinct as a result of forest conversion to agriculture. They were very heavily hunted for the fur trade until exports were banned in 1973–1981; some domestic trade in furs still occurs. Localised killing for furs, by dogs, in retaliation for killing poultry and as roadkill probably impacts populations close to people.

CITES Appendix I. Red List: Vulnerable. Population trend: Decreasing.

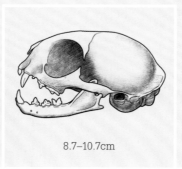

8.7–10.7cm

⬤ **IUCN RED LIST (2008):**
Near Threatened

Head-body length ♀ 47.7–62cm,
♂ 49–79.2cm
Tail 30–52cm
Weight ♀ 2.3–3.5kg, ♂ 2.3–4.9kg

Margay

Leopardus wiedii (Schinz, 1821)

Tree Ocelot

Taxonomy and phylogeny

The Margay is classified in the *Leopardus* genus (and lineage of the same name), and is most closely related to the Ocelot. The species shows high genetic heterogeneity, with three distinct population clusters: northern South America and southern South America, divided by the Amazon River, and Mesoamerica with weak genetic differences in a cline from Mexico to Panama. This would suggest three subspecies, although up to 10 are currently described; most are probably invalid.

Description

The Margay is the size of a lean domestic cat, with a slender, lightly built body and a very long, thickly furred tail with a tubular appearance. The front paws are hefty with very large, splayed digits. The head is compact and rounded with large ears and distinctive, very large eyes. Body fur is dense and soft, ranging in background colour through shades of greyish-buff, ochre, tawny and cinnamon-brown graduating to cream or white underparts. The Margay is richly marked with large, dark rosettes and blotches that usually coalesce into long stripes on the head, nape and back.

Similar species The Margay is easily confused with the Ocelot and oncillas; young kittens of all these species can be very difficult to tell apart. The Margay is significantly smaller than the Ocelot, though the two species overlap in weight at their respective extremes, and Margays are generally larger and more richly marked than oncillas. The Margay's most distinctive features are its very large eyes, proportionally very long tail and oversized front paws.

Distribution and habitat

The Margay occurs from Sinaloa and Tamaulipas, northern Mexico, throughout Central and South America to northern Argentina, eastern Paraguay and north-west Uruguay. One specimen collected along the Rio Grande near Eagle Pass, Texas, around 1850, is the only US record, possibly a pet or released individual given that the arid habitat is very atypical for the species. Margays are forest-dependent, and are more closely associated with forest habitats than any other neotropical cat. However, within that specialisation, they are able to occupy all kinds of evergreen and deciduous forests, from lowland tropical forest to montane cloud forests, from sea level typically to 1,500m, and exceptionally to 3,000m in the Andes. They occur in forested enclaves and gallery forest in dry forest-savanna mosaics, for example in Brazilian *caatinga* semi-arid thorny savanna and dry Uruguayan savanna. They also occupy secondary forest, forest-plantation mosaics and forest patches in highly disturbed landscapes. They tolerate converted landscapes provided these are densely vegetated, such as plantations of coffee, cocoa, eucalyptus and pine, but they cannot inhabit open agriculture including sugarcane, soy and pasture.

Feeding ecology

The Margay is a spectacularly acrobatic climber and is possibly the most arboreal of all felids with several anatomical adaptations for arborealism. The broad feet have very mobile digits with long, loosely knit metatarsals and enlarged claws, the hind ankles can rotate inwards through 180°, and the highly elongated, muscular tail assists with balance. They can descend head-first down trees, hang upside down by the hindfeet while handling objects with the forepaws, and scurry rapidly upside down along branches, all of which can be executed at high speed. Dietary records and direct observations are limited but they provide ample evidence of arboreal hunting including of very agile species such as small primates; however, most prey is terrestrial and Margays forage predominantly on the ground. Most prey weighs less than 200g. Typical prey includes small rodents such as Slender Harvest Mice, spiny

The Margay's very large eyes are distinctive, occupying a much greater proportion of the face than the similar Ocelot with which is it often confused (compare with the photo p.105). Additionally, Margay eyes are typically a few shades darker than the Ocelot's.

Margays are able to move rapidly along flexible vines and lianas including even the narrowest spans which they navigate by essentially running upside down at high speeds.

pocket mice, Short-tailed Cane Mice, Big-eared Climbing Rats, tree squirrels, as well as shrews and small marsupials such as mouse opossums. Larger prey includes Southern Opossums, cavies, agoutis, pacas and Brazilian Rabbits. Widespread reference to sloths, capuchins and tree porcupines as prey stems from a misidentified Ocelot; these species, especially juveniles, may be vulnerable to Margays though there is little evidence. A Margay in Guanacaste, Costa Rica, was disturbed feeding on a Mexican Tree Porcupine carcass, which it may have killed. Margays hunt tamarins and marmosets; captive individuals caught Red-handed Tamarins in semi-natural conditions, and a wild Margay was observed unsuccessfully hunting Pied Tamarins in Amazonia, Brazil. Local people there believe that Margays imitate the vocalisations of juvenile tamarins to attract adults, apparently confirmed by one recent scientific report, though its validity is questionable (local people also report Jaguars, Pumas and Ocelots imitating the calls of prey for which there is no evidence). Other common prey includes small passerine birds, cracids (guans and chachalacas),

Below: **There are very few direct observations of reproductive behaviour in wild Margays. Given the extent of the Margay's arborealism, it is entirely possible that courtship and mating behaviour occurs above ground in the canopy (C).**

lizards and frogs. A female Margay and her two kittens opportunistically raided bats captured in mist nets during survey work in Brazilian Atlantic Forest fragments, suggesting that bats are considered prey. Invertebrates are often consumed, though probably contribute little to intake, and fruit appears quite often in scat samples, possibly the stomach or crop contents of frugivorous prey. Captive Margays are said to readily eat figs. They kill domestic poultry, though depredation is limited to settlements near dense vegetation.

Margays forage mainly at night with peak activity around 21:00–05:00, though a single radio-collared male in Brazil showed cathemeral patterns. Margays are equally at home on the ground or in trees, but two radio-collared individuals (a female in Brazil and a male in Belize) both moved mainly on the ground, presumably locating both terrestrial and arboreal prey, and rested during the day in trees. Few hunts have been observed. In Uruguay, a Margay on the ground leapt 2m vertically to catch a perched guan, and a Margay in Brazilian Atlantic Forest spent 20 minutes unsuccessfully pursuing a bird in a bamboo stand to heights of 6m.

Social and spatial behaviour

Margays are poorly studied, with a handful of animals radio-collared in Belize, Brazil and Mexico. Margays are solitary and demarcate home ranges with characteristic felid marking behaviour. Range size for males is 1.2–6km² (El Cielo Biosphere Reserve, Mexico), 11km² (one subadult male, Belize) and 15.9km² (one subadult male, Brazil). Small range sizes at El Cielo may represent optimum conditions with its high-quality habitat, no large carnivores and good protection. Very few range estimates exist for females; one collared Brazilian female in a fragmented subtropical forest-agricultural mosaic used 20km². Range overlap between four radio-collared males in El Cielo was considerable, and the degree to which Margays defend ranges is unclear. Margays are capable of covering large distances rapidly. A subadult Belizean male (possibly dispersing) travelled up to 1.2km per hour and covered an average of 6.7km per day. He rested always above ground, 7–10m high in vine or liana tangles or in the boles of Cohune Palms, and moved every two to three hours during the day between rest sites, travelling on the ground. Based on camera-trapping, Margays occur at lower densities than Ocelots but there are few rigorous estimates; 12.1 individuals per 100km² was estimated for protected montane, pine-oak forest in central Mexico where Ocelots, Pumas and Jaguars are present (La Reserva Natural Sierra Nanchititla).

Reproduction and demography

Unknown from the wild. In captivity, the Margay has surprisingly low reproductive rates relative to its size, with protracted gestation and small litters. Reproduction is aseasonal. Gestation is 76–84 days. Unusually, female Margays have only one pair of teats and litter size is usually one kitten (very rarely two). Weaning occurs at around eight weeks, and female sexual maturity is 6–10 months but first litters in the wild are likely to be at two to three years.

Mortality Poorly known.

Lifespan 24 years in captivity, but certain to be much less in the wild.

Pairs of kittens are exceptional for Margays, with single births being most common. The pattern is also true for other members of the *Leopardus* lineage, particularly the Ocelot and oncillas.

STATUS AND THREATS

The Margay is strongly forest dependent. Although the species is sometimes recorded from highly disturbed habitat, including forest-agriculture mosaics in Brazil, it is thought to adapt poorly to forest conversion and fragmentation, which are the main threats. Outside extensive forest blocks, such as in the Amazon Basin, Margay populations are regarded as fragmented and declining in many areas. The species has surprisingly low reproductive potential which limits its ability to recover rapidly from declines or to recolonise former habitat. Additionally, its numbers are thought to be significantly influenced by the larger and much more adaptable Ocelot. The Margay was heavily hunted at the height of the fashion for spotted cat fur, with a minimum number of 125,547 skins legally exported from 1976 to 1985. Fortunately, all international trade in Margay fur is now illegal, though they are sometimes illegally hunted for domestic use. Persecution for killing poultry occurs, and kittens are sometimes captured as pets. Illegal trade and killing are likely to have significant effects on Margay populations in areas where there is already pressure on habitat.

CITES Appendix I. Red List: Near Threatened. Population trend: Decreasing.

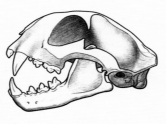

11.7–15.8cm

● **IUCN RED LIST (2008):**
Least Concern

Head-body length ♀ 69–90.9cm,
♂ 67.5–101.5cm
Tail 25.5–44.5cm
Weight ♀ 6.6–11.3kg, ♂ 7.0–18.6kg

Ocelot

Leopardus pardalis (Linnaeus, 1758)

Taxonomy and phylogeny

The Ocelot is classified in the *Leopardus* genus (and lineage of the same name), and is most closely related to the Margay. The species shows high genetic heterogeneity across its range, with four distinct population clusters: southern South America separated from all populations to the north by the Amazon River; northern South America which comprises two clusters – eastern Panama, north-west Brazil, Venezuela and Trinidad (and presumably

Colombia which was not sampled), and northern Brazil and French Guyana (Suriname and Guyana were not sampled); and Central America and Mexico. This would suggest four subspecies, although up to 10 are currently described; most are probably invalid.

Description

The Ocelot is the third largest cat in Latin America. The smallest adult females are about the weight of a large domestic cat. The Ocelot is robustly built with thickset limbs and a relatively short, tubular tail typically too short to reach the ground (but exceptions exist). The head is powerfully built with a blocky muzzle, especially in adult males, and rounded ears that are black with a white central spot on the back. The paws are heavily built, especially the forepaws that are much larger than the hindpaws (giving rise to various local names such as *manigordo* or 'fat hand'). The body fur is dense and soft, and very variable in background colour within and between populations, including shades of creamy-buff, tawny, cinnamon, red-brown and grey, with white underparts. Ocelots are very richly marked with highly variable combinations of black open and solid blotches, streaks and rosettes with russet-brown centres. Simple solid spots or blotches usually cover the lower legs, and the tail has black partial or complete rings with a black tip.

Similar species The Ocelot is very similar to the Margay but usually considerably larger, more heavily built and with a distinctively shorter tail. Young kittens of Ocelot, Margay and oncillas can be very difficult to tell apart.

All cats have highly sensitive whiskers or vibrissae that assist with moving in near or total darkness and also help them react instantly to the movements of prey at the moment of the killing bite. Cats also possess a second type of long tactile hair called a tylotrich scattered thinly over the body in among the less sensitive guard hairs.

Distribution and habitat

The Ocelot occurs from northern Mexico throughout Central and South America to south-eastern Brazil and northern Argentina including on Trinidad (with recent camera-trap confirmation from 2014) and Isla de Margarita, Venezuela. It does not occur in Chile and its presence in Uruguay is now uncertain although it occurs in Brazil very close to the border. The Ocelot formerly occurred in the southern US from Arizona to Arkansas and Louisiana but the only remaining breeding populations occur in two isolated fragments (Laguna Atascosa National Wildlife Refuge and Willacy County) in extreme southern Texas, numbering a total of 60–100 animals. Between 2009 and 2013, at least four male Ocelots were photographed in the Huachuca Mountains, Santa Rita Mountains and Cochise County of southern Arizona. Ocelots occur in a wide range of habitats including dense thorn-scrub, shrub woodlands, wooded savanna grasslands, mangroves, swamp-woodland mosaics and all kinds of dry and moist forest, but they strongly prefer dense cover within all habitat types and are not nearly as generalist as comparable species, for example the Bobcat. They are tolerant of modified habitat provided there is dense vegetation and prey; they occur in secondary forest and agricultural landscapes with extensive brush such as fallow cultivation. They mostly avoid very open areas but readily hunt in pasture and grasslands close to cover, especially at night. Ocelots are usually found between sea level and 1,200m, rarely up to 3,000m.

Feeding ecology

The Ocelot is muscularly built with large, very powerful forepaws and a robust skull with large canines, a prominent sagittal crest and strong zygomatic arches conferring great biting power.

The size of Ocelots differs across the range although not along a north–south cline as often reported. Habitat type is the main determinant of differences, with rainforest cats being the largest and those from dry habitats such as scrub, chaparral and dry forest being the smallest.

These adaptations enable the Ocelot to overpower large prey including sloths, tamanduas, howler monkeys, juvenile Collared Peccaries and juvenile White-tailed Deer; there are also records of Ocelots feeding on adult Red Brocket Deer, though it is unclear if they were scavenged. Despite this, and based on early studies, the Ocelot has long been regarded as subsisting principally on very small mammals weighing less than 600g. A more accurate picture of the Ocelot's diet lies somewhere between these extremes. For many populations, the most important prey category comprises medium-sized vertebrates, mainly large rodents such as acouchis, agoutis and pacas, opossums, armadillos and, in some cases, large reptiles particularly Green Iguanas (~3kg). This is supplemented by frequent kills of very small rodents which, although eaten often, may contribute less to intake. Ocelot diet is flexible and the relative importance of these prey categories

varies depending on the region and availability of species. The largest prey (on average) killed is by Ocelots on Barro Colorado Island, Panama, where they eat mainly Hoffmann's Two-toed Sloth, Brown-throated Sloth, and agoutis (33 per cent of their diet compared to a range of 0 per cent in Brazilian Atlantic Forests to 10.7 per cent in Iguaçu National Park, Brazil) as well as pacas, Green Iguanas and White-nosed Coatis.

Other relatively common prey includes squirrels, rabbits, cavies, tree porcupines and small primates, such as tamarins and squirrel monkeys, as well as birds including tinamous, guans, woodpeckers and doves. Ocelots readily consume aquatic and semi-aquatic prey including fish, amphibians and crustaceans, indicative of their ability to live in inundated habitats. Ocelots are highly opportunistic and shift prey preferences depending on availability; for example, large land crabs are heavily consumed

Below, left and right: **Cats are wonderfully attuned to hunting possibilities as they move through their environment. Uncertain of a possible target, this Ocelot raises itself on its back legs and silently scales a nearby tree to gather more information before deciding on the hunt.**

when they become abundant during the wet season in Venezuelan *llanos* (flooded savanna). Incidental prey includes virtually any small species encountered, with records of bats, lizards, snakes, small turtles, caimans (presumably hatchlings) and arthropods. Relatively few carnivores are recorded in the diet – coatis and Crab-eating Raccoons are the most common carnivores killed, otherwise there are single records of olingo, Kinkajou, Crab-eating Fox, Tayra, Lesser Grison, Margay and oncilla as prey. Reptile and bird eggs are opportunistically eaten. There are no records of cannibalism except very rarely in captivity. Ocelots sometimes kill poultry but otherwise are not recorded preying on domestic animals.

Ocelots have been observed moving at all times of day but camera-trapping clearly shows they are primarily nocturno-crepuscular, with most activity taking place between 20:00 and 06:00. Diurnal activity increases on overcast, cool days and during the wet season (e.g. in Venezuelan *llanos*, which has

more cloudy days). A Peruvian radio-collared mother with a kitten increased her diurnal foraging so that she was active for 17 hours a day, presumably to feed them both. Direct observations of hunting are surprisingly rare given Ocelots are widespread and often the most abundant felid present. Hunting is chiefly terrestrial, though they are adept climbers and the relatively high frequency of arboreal species in the diet suggests some prey is taken in trees – there is one record of an Ocelot attacking a young Mantled Howler Monkey in a tree (Barro Colorado Island). The Ocelot appears to adopt two main hunting techniques: scanning and listening for prey while quietly walking, interspersed with periods of sit-and-wait ambush hunting. In the latter technique, they may sit for up to an hour, often on an elevated point such as a fallen tree, to wait for prey to give away its location. Although Ocelots hunt mostly in or near dense cover, they frequently search for prey along trails, on river or coastal beaches and at the edge of open areas where prey is abundant and easily detected. The Ocelot is an excellent swimmer and frequently traverses large rivers; while there is no evidence for hunting in deep water, Ocelots forage in shallow water at the edges of water bodies and in sodden habitats such as seasonally flooded savannas (e.g. *llanos* and Pantanal). Ocelots kill most small to medium-sized prey by biting the nape or skull; for example, six Central American Agoutis weighing 2–3kg were killed by a skull bite (Barro Colorado Island).

Ocelots readily scavenge, for example from refuse piles left by people fishing in the southern Pantanal, Brazil, where Ocelots visit known dumpsites nightly to eat fish offal. They frequently cache food by covering it entirely with leaf and soil debris, and may return to feed on kills covered in this way over a number of nights.

Social and spatial behaviour

The Ocelot is solitary, with a typical felid socio-spatial pattern where studied. Adults maintain

An Ocelot with captured Cobalt-winged Parakeet it has batted from the air at a clay lick where the parrots congregate for minerals, Tambopata National Reserve, Peru.

Like all cats, Ocelots leave information for conspecifics through a variety of territorial markings from claw raking, urination and defecation. Ocelots are known to occasionally use communal latrines including, in one case, the concrete floor of a small observation shelter.

enduring home ranges, usually with exclusive core areas and considerable overlap at the edges. Male ranges are generally double to four times the size of female ranges, and are less exclusive with greater overlap at the edges. Adults patrol and scent-mark ranges that are defended against same-sex conspecifics in sometimes fatal fights. Resident adults of both sexes are tolerant of their offspring, which may linger in the natal range for more than a year after independence, with some evidence of regular interaction. Like most felids, familiar Ocelots are likely to interact with each other regularly even if most hunting and other daily activities are solitary. Unfamiliar immigrants risk aggression from residents, especially males, which are responsible for

the deaths of some dispersers. For example, two transients seeking territory were killed by resident males in Texas (comprising 7 per cent of 29 deaths between 1983 and 2002).

Estimates of range size by radio-telemetry have been conducted in Argentina, Belize, Brazil, Mexico, Peru, Venezuela and the US. Territories are quite small compared to similar-sized felids; the smallest ranges are in seasonally flooded savannas (Brazilian Pantanal and Venezuelan *llanos*), and the largest in dry *cerrado* savannas (Emas National Park, Brazil). Male ranges (average 5.2–90.5km²) overlap multiple, smaller female ranges (average 1.3–75km²). Ocelots appear to reach higher densities than all smaller felids in virtually all cases in which they have been

studied, and reach very high numbers in good habitat. Density estimates include 2.3–3.8 per 100km² (tropical pine forest, Belize), 13–19 per 100km² (Atlantic Forest, Brazil), 26 per 100km² (tropical rainforest, Belize) and 52 per 100km² (dry Chaco-Chiquitano forest, Bolivia).

Reproduction and demography

The Ocelot has unexpectedly low reproductive rates, with fairly long gestation, very small litters and long inter-litter intervals such that the lifetime reproductive output per female is only about five young, similar to (and, in some cases, fewer than) the output of large felids. Breeding is aseasonal. Gestation is 79–82 days. Litter size is one to two, with a single kitten being most common in captivity – litters of three have been recorded exceptionally in captivity. Kittens are slow to mature and reach independence at 17–22 months, and they may stay in the natal range until two to three years old. Dispersal appears to follow a typical felid pattern with males dispersing farther than females, though data (especially for females) are limited, and most monitored dispersers died before settling. Dispersal distances range from 2.5 to 30km. Six surviving Texan dispersers (of nine monitored) settled 2.5–9km from their natal ranges.

Mortality Rates are unquantified for most populations. Average annual mortality in Texan Ocelots ranges from 8 per cent for resident adults to 47 per cent for dispersing subadults; 45 per cent of mortality in this population is anthropogenic, primarily from roadkills, compared to 35 per cent from natural causes (the balance due to unknown causes). Known predators include Jaguars, Pumas and domestic dogs. There is one record each of an adult killed by a Boa Constrictor and an American Alligator (Texas). Ocelots are presumably vulnerable to large anacondas, caimans and perhaps Harpy Eagles. One collared Ocelot in Texas died after a rattlesnake bite.

Lifespan Unknown in the wild, 20 years in captivity.

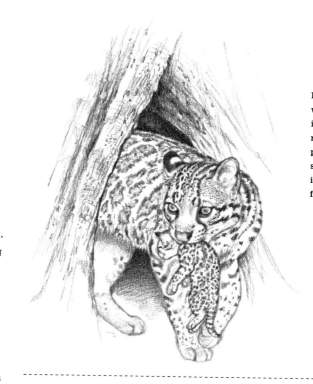

Despite the Ocelot being widespread and common in many areas, its reproductive ecology is poorly understood. The survival rate of kittens is essentially unknown from the wild.

STATUS AND THREATS

The Ocelot is widespread, occupies a wide range of habitats and is able to reach high densities. It is considered secure and common in much of its range. Range loss is prevalent at the extremes of its historical distribution where it is now extinct or relict, especially in the southern US, northern and western Mexico, much of eastern and south-eastern Brazil, Uruguay and northern Argentina. Through much of the remaining, current range, the species occupies extensive areas of continuous or near-continuous distribution.

Nevertheless, the Ocelot depends on dense habitat and has very low reproductive potential, making it vulnerable to human pressures, and it is poorly equipped to recover or recolonise after threats subside. Habitat destruction is the main driver of declines, together with illegal hunting. Ocelots were once intensively hunted to fill the demand for spotted fur, with 140,000–200,000 skins exported annually from Latin America during the 1960s and 1970s. International trade became illegal in 1989 and hunting is prohibited by all range states, but hunting is still fairly widespread, for recreation and to sell furs into illegal domestic trade and international smuggling. Ocelots are easily hunted by treeing with dogs, a common practice across much of the range especially in ranching landscapes. Ocelots are also often killed intentionally or as bycatch in retaliation for poultry depredation. Humans may compete for Ocelot prey as key prey species, such as pacas, agoutis and armadillos, are very widely hunted by people throughout the range, and the two most important prey species for Ocelots in Jalisco, Mexico – White-tailed Deer and Spiny-tailed Iguana – are hunted heavily for human consumption.

CITES Appendix I. Red List: Least Concern. Population trend: Decreasing.

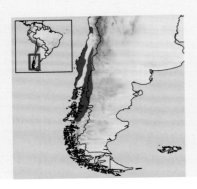

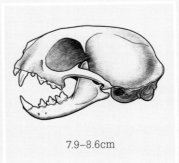

7.9–8.6cm

⬤ **IUCN RED LIST (2015):**
Vulnerable

Head-body length ♀ 37.4–51cm,
♂ 41.8–49cm
Tail 19.5–25cm
Weight ♀ 1.3–2.1kg, ♂ 1.7–3kg

Guiña

Leopardus guigna (Molina, 1782)
Guigna, Kodkod, Chilean Cat

Taxonomy and phylogeny

The Guiña is classified in the *Leopardus* genus and is most closely related to Geoffroy's Cat with a shared ancestor estimated at less than a million years ago. Both species were formerly grouped in their own genus *Oncifelis* with the Colocolo. The most recent consensus is the two species clearly represent a sub-branch of the *Leopardus* genus together with the oncillas to which both are related. Two Guiña subspecies are described, with moderate morphological and genetic differences. *L. g. guigna* occurs in southern Chile and is apparently slightly smaller and more brightly coloured than the paler *L. g. tigrillo* of central Chile.

Description

The Guiña is the smallest cat in the Americas. It is a tiny, compactly built species with relatively short limbs and a thick, tubular bushy tail. The fur is greyish-brown to rich russet-brown marked with small, dark dab-like spots that coalesce into broken lines on the back and nape. The head is small and rounded, with a compact face that is distinctively marked with dark cheek stripes, eyebrow markings and prominent dark stripes under the eyes that border the muzzle; this gives the Guiña's face a superficial resemblance to a Puma kitten. Melanism is common, sometimes with mahogany brown (rather than black) extremities on which the markings are obvious. At two sites in southern Chile 16 of 24 captured individuals were black.

Left: **A melanistic Guiña kitten in the Parque Nacional Laguna San Rafael in Chile. The Laguna San Rafael population occurs on the Taitao Peninsula at the southernmost limit of the species' range, isolated from other populations by extensive ice fields to the east.**

Below: **This captive adult female Guiña is the larger, more lightly coloured northern subspecies *L. g. tigrillo* that occurs in Central Chile (C).**

Similar species The Guiña is very similar in appearance to the closely related Geoffroy's Cat, which is larger with a relatively heavier head and less bushy tail. The two species are known to overlap only at the extreme eastern edge of the Guiña's range, for example in Los Alerces National Park, southern Argentina and Puyehue National Park, Chile.

Distribution and habitat

The Guiña has the smallest distribution of any Latin American cat, restricted to central and southern Chile including Isla Grande de Chiloé and marginally in adjacent border areas of extreme south-west Argentina. Guiñas are strongly associated with dense, temperate rainforest and southern beech forest, particularly the distinctive Valdivian forests in southern Chile characterised by dense *colihue* bamboo thickets and fern understorey; in central Chile, they occur mostly in Chilean *matorral* habitat made up of temperate forest, woodland and dense scrub. They occur in suitable habitat from sea level to the treeline at 1,900–2,500m. Guiñas consistently avoid open land such as cultivation and areas with short vegetation except for very small patches, which they traverse to reach cover. They inhabit secondary forest, forested ravines and coastal forest strips in heavily altered habitat, and they use small forest fragments and plantations, for example of eucalyptus and pine, when close to native forested habitat, and provided it contains dense understorey.

Feeding ecology

Guiñas hunt mainly very small rodents especially Olive Grass Mice, Long-haired Grass Mice, Chilean Climbing Mice and Long-tailed Pygmy Rice Rats; very small marsupials (e.g. the 16–42g Monito del Monte); and adult and nestling birds, especially of

A wild Guiña in Valdivian rainforest of the La Araucanía region in southern Chile. This is the slightly smaller and more richly coloured southern form, *L. g. guigna*.

'flutterers' (near-flightless species) such as Chucao Tapaculos and huet-huets, and ground-foraging or nesting species such as Thorn-tailed Rayaditos, Austral Thrushes and Southern Lapwings. Small reptiles and insects are also consumed, though they probably contribute relatively little to intake. They kill domestic chickens and geese in fragmented, human-dominated landscapes (e.g. on Chiloé), especially where poultry is free-ranging but also by raiding hen houses; domestic goose is the largest known prey species, though they are not recorded preying on comparatively sized wild bird species. Reports by local people of the species killing goats and hunting in groups numbering up to 20 are implausible.

Foraging appears to be cathemeral with a tendency to be most active at dusk into the evening. Hunting mostly takes place on the ground and into the lower understorey in areas of very dense

undergrowth where small mammals and birds are most common. They are capable climbers and actively hunt in lower branches. They have been seen stalking small lizards in trees growing in steep-sided ravines and have recently been photographed raiding artificial nest-boxes in trees (up to 1.5m) used by cavity-nesting birds and small mammals. Nest-predation is probably an important source of prey for this species. Although fish has not been recorded in the diet, they readily swim and a young male was observed unsuccessfully attempting to catch fish in tidal pools for 10 minutes; it is likely that fish and marine invertebrates are occasionally eaten. Guiñas are known to scavenge, for example from Pudú and sheep carcasses, though the availability of large carrion in their environment is limited.

Social and spatial behaviour

Guiñas are essentially solitary and appear to establish small, stable home ranges, but range overlap between adults and independent subadults can be considerable. During a 3.5-year study of two

Above: **A Guiña pauses on a tarred tourist road in Parque Nacional Puyehue, Chile. Road construction is a factor contributing to the larger, pervasive threat of habitat fragmentation, which has led to many Guiña populations being isolated in forest islands.**

Left: **Monitoring of artificial nest-boxes has revealed the ability of the Guiña to fish for nestlings, behaviour which almost certainly extends to natural tree hollows.**

protected populations (Laguna San Rafael and Queulat national parks) in southern Chile, all radio-collared individuals overlapped extensively with all their neighbours. Shared use included the core areas of ranges with very little evidence of territorial behaviour. Aggressive interactions appear to be rare and mild. In the same study, only two aggressive encounters were observed, both between male pairs which were resolved without injury; in both cases, the younger male remained in the vicinity, suggesting fights do not play a strong role in evicting same-sexed rivals from territories, as occurs in many other felids. Territoriality may be more prevalent among populations on Chiloé where core areas were somewhat more exclusive for a small number of adults, and there was limited evidence of two adult males patrolling their range boundaries with constant loud vocalising at the borders. The Chiloé cats live in anthropogenically modified habitat in which the availability of prey may be more limited, leading to greater territoriality.

In protected populations, range size is similar for males and females, averaging 2.4 km^2 (females) and 2.9km^2 (males). Females in fragmented habitat have somewhat smaller ranges than males, though this is based on a very small number of animals; for example, 0.6km–1.7km^2 for females, compared to 1.6–3.7km^2 for males on Chiloé; and 1.2–3.2km^2 for females compared to 2.3–4.8km^2 for males in human-modified habitat in the Araucanía region. In a 24-hour period, Guiñas may travel up to 9km, with no differences between the sexes, on average 4.5km for females, compared to 4.2km for males. Density estimates from radio-telemetry counting adults and subadults are high: 1–3.3 cats per km^2.

Reproduction and demography

This is largely unknown from the wild. Guiñas experience very cold winters which possibly drives seasonal breeding, though this is poorly known. Based on the estimated age of a small number of captured kittens in southern Chile, a possible mating season period occurs in the early spring, August to September, with births in late October to early November. In captivity, gestation is 72–78 days and litter size is one to three.

Mortality Poorly known. The Puma is the largest potential predator, though they are rare or transient in most of the Guiña's current range. The large Culpeo Fox is a potential predator of kittens. Where studied, people and their dogs are the main cause of death, for example two of seven radio-collared cats on Chiloé.

Lifespan Unknown in the wild, 11 years in captivity.

STATUS AND THREATS

Guiñas have a very restricted distribution (about 300,000km^2) and are closely tied to unique, dense temperate forest habitat that is highly threatened. Clearing of forest for agriculture and plantations has reduced Guiña range to many small, fragmented populations. Fragmentation is severe in central Chile where there are an estimated 24 isolated subpopulations, 90 per cent of which contain fewer than 70 individuals. Southern Chile is more of a stronghold, with lower human densities and several large, protected areas with Guiña populations. Guiñas are widely persecuted as poultry pests, though actual damage appears to be less than believed; for example, only 4.5 per cent of 199 chicken owners interviewed in the Araucanía region reported losses to Guiñas in the preceding year (2011). Nonetheless, they are readily killed illegally by local people, facilitated by their habit of fleeing to trees when pursued. They are also killed incidentally during legal fox hunts by people using dogs and traps. Due to their small size and relatively drab fur, Guiñas have never been especially targeted for their skins.

CITES Appendix I. Red List: Vulnerable. Population trend: Decreasing.

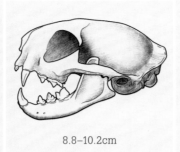

8.8–10.2cm

● **IUCN RED LIST (2008):** Least Concern

Head-body length 42.3–79cm
Tail 23–33cm
Weight 1.7–3.7kg

Colocolo

Leopardus colocolo (Molina, 1782)

Pampas Cat, Pantanal Cat

Colocolo form

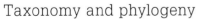
Pampas form

Pantanal form

Taxonomy and phylogeny

The Colocolo has been classified alternatively as a member of the now invalid genus *Oncifelis* with the Guiña and Geoffroy's Cat, or in its own unique genus *Lynchailurus*, but recent molecular analyses place it firmly within the *Leopardus* genus. Its closest relative is thought to be the Andean Cat, though the genetic data are poor and further research is required. The Northern Oncilla is also closely related, with evidence of historical hybridisation between Northern Oncilla males and female Colocolos in north-east and central Brazil (the two species are not thought to presently hybridise although this possibility exists and may be revealed by ongoing research). Based on morphology, the Colocolo clusters into three major groups (see Description) that have sometimes been classified as full species, but available genetic data indicate only moderate differences between populations. Up to eight subspecies are described but there is little molecular support for so many; this is likely to be revised with better data.

Description

The Colocolo is relatively small and stocky, about the size of a domestic cat with relatively short legs and a shortish, thickly furred tail. In contrast to the rounded ears of all other small felids in South America, the ears are distinctly triangular. The fur colour and patterning is highly variable, with three distinct morphs in which the only constant is dark

brown-black striping on the forelegs. The morphs are broadly specific to a region, though there is variation within each and intergrading between them: 'Colocolo', with overall buff-grey colouration and marked with rich ginger blotches all over the body and dark tail bands, (High Andes of Peru, Chile and Argentina); 'Pampas Cat', the palest form with frosted grey fur lightly marked with large indistinct cinnamon blotches and darkish tail bands (Colombia to Patagonia on the eastern side of the Andes); and 'Pantanal Cat', the darkest form with rust-brown to mahogany-brown body fur lacking markings or lightly marked with indistinct blotches, and dark brown-black socks (eastern Bolivia, Brazil, Paraguay and Uruguay). Melanism is recorded from Brazil and Peru (and probably occurs more widely).

Similar species The 'Colocolo' form is very similar to the Andean Cat and is the most common morph occurring in Andean Cat range. The Colocolo is generally more heavily marked, lacks the distinct long, bushy tail of the Andean Cat and has a pink or reddish nose compared to the Andean Cat's dark nose.

Distribution and habitat

The Colocolo occurs along almost the entire length of the Andes from extreme southern Colombia to the Strait of Magellan, Chile, and inland throughout the Bolivian lowlands into central and northern Brazil, Paraguay, Uruguay and much of Argentina except most of the northern and central-eastern provinces. The range is often presented as a series of discontinuous populations but this likely reflects gaps in sampling rather than in distribution; ongoing camera-trapping efforts often document the species within former gaps. Colocolos inhabit a very broad range of habitat types; they are generally associated with open habitats, though they occur in some dense woodlands and forests. They are found in *pampas* and *cerrado* grasslands, all kinds of moist and dry savanna woodlands, marshland, mangroves, open forest, semi-arid scrub and desert, and Andean steppes to 5,000m. They are recorded from the dense

Yungas montane forest belt along the eastern slopes of the Andes between the Andean highlands and eastern lowlands; they do not occur in lowland tropical and temperate rainforest. Colocolos occur in some highly modified habitats including rangelands, exotic plantations and agricultural landscapes provided there is dense ground cover present. They do not tolerate highly degraded pasture or extensive crop monocultures, which includes most of the present-day *pampas* grasslands in Argentina.

Feeding ecology

The Colocolo is poorly studied; radio-telemetry studies are underway in the Argentinean Andes and Brazilian *cerrado* grasslands that are certain to provide a better understanding of this species' ecology, but the data are not yet available. Information on Colocolo feeding ecology comes mainly from analysis of scats and stomach contents. This shows the species to be an opportunistic generalist that preys mainly on small mammals. Rodents are the primary prey including leaf-eared mice, field mice, rats, cavies, tuco-tucos, chinchilla rats and mountain viscachas. In Andean areas, the

The Colocolo is sufficiently powerful to kill neonate ungulates such as Guanaco lambs although there are no confirmed records. Most observations of Colocolos feeding on ungulate carcasses are assumed to be cases of scavenging.

Opposite: **The Colocolo was once the most heavily traded species during the peak of the fad for spotted cat furs from Latin America. More than 78,000 Colocolo skins were exported between 1976 and 1979 from Argentina alone.**

Below: **Surprised in the open by the observer, this Colocolo adopts a common defensive tactic of small felids in sparse habitat, of sinking close to the ground and freezing, hoping to avoid detection. San Pedro de Atacama, Chile.**

most important prey is viscacha, overlapping significantly with the diet of the Andean Cat; however, Colocolos are the more versatile species, with a broader diet. Introduced European Hares are important prey in Patagonia. Other notable prey includes tinamous, adult flamingos and Magellanic Penguin chicks and eggs, which they take in the nest. Small lizards and snakes are recorded, as well as invertebrates. They kill domestic poultry. Based on diet records, foraging is assumed to be largely terrestrial, though Colocolos are very adept climbers and they likely pursue prey arboreally, at least within lower branches. Activity periods appear to vary considerably with region. Based on camera-trap results, they are largely nocturnal in the Argentinean Andes while they are almost entirely diurnal in Brazilian *cerrado* grasslands. The presence of nocturnal large cats in the *cerrado* is suggested as a factor, and yet Pumas are common in

Andean habitats where both species are largely nocturnal. Like many small neotropical felids, they probably have flexible activity patterns which they adjust depending on a variety of factors. Colocolos scavenge from large carcasses, including those of livestock, Vicuña and Guanaco.

Social and spatial behaviour

The Colocolo is solitary and assumed to have a typical small felid socio-spatial system but few specifics are available until the results of ongoing telemetry studies are published. Preliminary data from radio-tracked animals estimates the home range size as 3.1–37km^2, averaging 19.5km^2 (*cerrado* grasslands, Emas National Park, Brazil), and a single estimate of 20.8km^2 (Argentinean Andes); ranges in the Andes are likely to be larger given that prey is patchily distributed in lower densities than in *cerrado* but data from more animals are required.

Similar to some other small neotropical felids, Colocolos deposit faeces in latrines that probably have a spacing and 'signpost' function. There are few rigorous density estimates available; based on camera-trap data, they have been estimated at 74–79 cats per 100km² in the Argentinean High Andes.

Reproduction and demography

This is largely unknown from the wild. Based on limited information from captivity, gestation lasts 80–85 days and the litter size is one to three kittens; the average size from 13 captive litters was 1.31.

Mortality Poorly known. Larger cats presumably kill them occasionally; there is one record of Puma predation from Patagonia. Domestic dogs kill them frequently in some locations, for example in open habitat, north-west Argentina.

Lifespan Unknown in the wild, up to 16.5 years in captivity.

STATUS AND THREATS

Colocolos are widely distributed with a broad habitat tolerance and have historically been regarded as common. Although the species is locally common at some sites, increased survey effort over the last decade indicates they are actually scarce across much of their range. In lowland areas, habitat loss and conversion for livestock and agriculture are the main threats; this is particularly pervasive in key habitats for the species such as Brazil's *cerrado* grasslands, Argentine Espinal and Gran Chaco woodlands. Agricultural expansion, especially for soybean, appears to be driving declines in central Argentina, and the species is now considered regionally extinct in the Argentinean *pampas* grasslands for which it was named. Colocolos are often killed on roads, are persecuted for raiding poultry and are vulnerable to shepherds' dogs, especially in open habitats. In Andean areas, Colocolos and Andean Cats are killed for religious ceremonial uses in which the skin or stuffed cat is believed to confer fertility and productivity on domestic livestock and crops.

CITES Appendix I. Red List: Least Concern. Population trend: Decreasing.

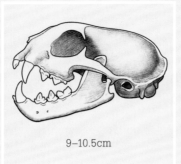

9–10.5cm

● **IUCN RED LIST (2008):**
Endangered

Head-body length 57.7–85cm
Tail 41–48cm
Weight 4.0kg (single ♂)

Andean Cat

Leopardus jacobita (Cornalia, 1865)

Andean Mountain Cat

Taxonomy and phylogeny

The Andean Cat was formerly classified in its own genus *Oreailurus*, but genetic analysis places it firmly in the *Leopardus* lineage. Its closest relative is thought to be the Colocolo but the evidence is equivocal and further research is needed. No subspecies are described but recent genetic analysis shows the southernmost population in Patagonia is an evolutionarily significant unit distinct from other populations.

Description

The Andean Cat is about the size of a very large domestic cat with a stocky build, thickset limbs and large paws. The face is lightly marked, with the exception of broad dark eye stripes running under the temples, and dark cheek stripes. The face has a distinct, slightly anxious expression. The tail is long with very thick fur giving it a bushy, tubular appearance. The thick fur is pale silver-grey marked with large russet blotches on the body that darken to rich grey-brown on the face, chest and limbs. The tail has 5–10 distinctive thick russet bands that become paired dark brown rings, often with a russet-brown centre, towards the tip.

Similar species The Andean Cat is easily confused with the 'colocolo' form of the Colocolo, which is sympatric over most of the Andean Cat's range. The Colocolo is generally more heavily marked, lacks the distinct long, bushy tail of the Andean Cat and has a pink or reddish nose compared to the Andean Cat's dark nose.

Distribution and habitat

The Andean Cat occurs in central and southern Peru, western Bolivia, north-east Chile and western Argentina, where it has a very restricted range in high Andean habitats mostly from 3,000 to 5,100m, with most records above 4,000m. In the Andes, they are restricted to semi-arid to arid sparsely vegetated areas above the timber-line, primarily in habitats dominated by rocky steep slopes with *bofedales* (well-watered shrub-grasslands supplied by glacial

A female Andean Cat and her kitten photographed by camera-trap on the Chilean High Plateau. Kittens are often slightly darker and more richly coloured than adults, giving young animals a stronger resemblance to the Colocolo.

Right: **Mountain viscachas and very small rodents make up most of the Andean Cat's diet. Andean Cats often consume many more mice than anything else, but at 25 times a mouse's weight, the viscacha is essential for the cat's survival.**

Below: **A wild Andean Cat searches for prey along arid, salty lake shore habitat at the Salar de Surire Natural Monument in the Chilean Andes.**

melt water) and associated dry scrublands. Andean Cats have also been recently recorded from Patagonian steppe in south-west Argentina at elevations of 650–1,800m; this is below the tree-line, in rocky areas with scrub and steppe vegetation.

Feeding ecology

The Andean Cat is highly dependent on mid-sized rodents living in rocky habitats. Its range overlaps the former distribution of the Short-tailed Chinchilla, which was probably a key prey species but is now reduced to a few tiny, Critically Endangered populations due to overhunting for fur. Andean Cats now specialise largely on two species of mountain viscachas (*Lagidium* spp.) that also share a very similar distribution. Mice, chinchilla rats, cavies, European Hares and tinamous are also important prey. They sometimes scavenge from carcasses of dead ungulates. Their impact on domestic species is poorly known. There is little evidence for them taking domestic poultry, which are rarely kept in most Andean Cat habitat. Small stock herders in Argentinean Patagonia give compelling eyewitness accounts of them taking down very young goat kids.

Based on limited camera-trapping and one female that was briefly radio-collared, Andean Cats appear to be mostly nocturnal with crepuscular activity peaks. Dusk and dawn are probably important hunting periods to coincide with activity peaks of viscachas. Sightings and activity data from telemetry during the day suggest some flexibility in foraging patterns. Andean Cats rarely occur in anything other than poorly vegetated habitats, and their foraging is ground-based, typically in very uneven, rocky habitat in which they are adept at hunting at high speed.

Social and spatial behaviour

Poorly known. Most sightings and camera-trap records indicate largely solitary behaviour. Only six animals have ever been radio-collared, and data from only one animal, a female in the Bolivian Andes, has been published. She was studied for seven months,

and her home range was estimated at 65.5km^2, which is unexpectedly large. Preliminary data from five collared cats in the Argentinean Andes suggest comparably large range sizes averaging 58.5km^2, which is twice the size of Colocolo ranges in the same area. The Andean Cat's large home ranges may reflect the species' reliance on viscachas that live in widely spaced colonies, so that cats may need to cover large distances moving from colony to colony. Andean Cats are always less abundant in camera-trap surveys than the closely related Colocolo where the two species are sympatric. The only rigorous density estimate for Andean Cats is from the Argentinean High Andes where they were estimated at 7–12 cats per 100km^2 compared to 74–79 Colocolos per 100km^2.

Recent surveys and occasional sightings by wildlife photographers have expanded the known range of the Andean Cat. Even with this, its total distribution is very limited and it is not common anywhere.

Reproduction and demography

This is largely unknown from the wild. Andean Cats experience very cold winters that are likely to drive seasonal breeding. The southern spring and summer from October to March is the probable breeding period and would coincide with the birth flushes of many prey species. Kittens have been observed between October and April. The gestation period is unknown but thought to be approximately 60 days, and litters are thought to number one to two kittens. There are no Andean Cats in captivity.

Mortality Poorly known. The Puma is the largest potential predator – they are rare or transient in much of the Andean Cat's range but abundant in the southern half of their range. Where studied, people and their dogs are the main cause of death.

Lifespan Unknown.

The Andean Cat has very low genetic diversity, comparable to cat species which have undergone population bottlenecks such as the Iberian Lynx and Cheetah. It is suggested that Andean Cats may have undergone similar declines during warm, interglacial periods when suitable high-altitude habitat receded.

STATUS AND THREATS

Andean Cat status is difficult to assess. During surveys, this species is found far less often than other carnivores suggesting it is naturally rare. They have a very restricted distribution and narrow habitat preference that is vulnerable to livestock grazing, agriculture, and mining and oil development, which often affect rocky habitats and water sources. Habitat conversion may also impact prey numbers, especially combined with human hunting of prey species, particularly viscachas, which is considered a serious threat. Andean Cats are sacred according to indigenous Aymara and Quechua 'harvest festival' traditions in which the skin or stuffed cat is kept in the house in the belief it confers fertility and productivity on domestic livestock and crops. They apparently have little fear of people and are very approachable when seen. They are easily killed by local people who simply stone them with a large rock. They are also persecuted for suspected poultry and livestock killing and are readily killed by goat-herders and their dogs in Patagonia, Argentina. Rapid and extensive development of 'fracking' in Argentina's northern Patagonian steppe threatens the entire Argentinean Patagonian range of the species. CITES Appendix I. Red List: Endangered. Population trend: Decreasing.

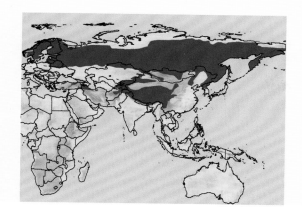

Head-body length ♀ 85–130cm,
♂ 76–148cm
Tail 12–24cm
Weight ♀ 13.0–21kg, ♂ 11.7–29.0kg

Eurasian Lynx

Lynx lynx (Linnaeus, 1758)

13.1–15.3cm

Taxonomy and phylogeny

The Eurasian Lynx is classified in the *Lynx* lineage and is closely related to both the Canada Lynx and Iberian Lynx. Some authorities formerly classified the Eurasian Lynx and Canada Lynx together as one species but genetic analysis confirms they are distinct with a common ancestor around 1.5 million years ago. As many as nine Eurasian Lynx subspecies are described. There are some genetic differences between European populations, suggesting separation into three subspecies: Carpathian Lynx *L. l. carpathicus* in the Carpathian Mountains (and most western European populations, which were reintroduced); Balkan Lynx *L. l. balcanicus*, restricted to the south-western Balkans, mainly the border regions of Albania and Macedonia;

and Northern Lynx *L. l. lynx* in the rest of Europe, including Fennoscandia, the Baltic States, north-eastern Poland and western Russia to the Yenisei River, and central Siberia. Differentiation between populations across their vast Russo-Asiatic range is unclear, but historically, the following subspecies are described: *L. l. dinniki* (Turkey, Caucasus and northern Iran); *L. l. isabellinus* (central Asia); *L. l. wardi* (Altai Mountains); *L. l. kozlovi* (northern Mongolia and southern Siberia); *L. l. wrangeli* (eastern Siberia, east of the Yenisei River); and *L. l. stroganovi* (Russian Far East).

Description

The Eurasian Lynx is the largest lynx species by a considerable margin, weighing twice as much on average as the Canada Lynx or Bobcat. Like all lynx, the species has a relatively lightly built body with elongated legs and large feet giving it a tall, leggy appearance, and a very short, black-tipped tail. The hindlimbs are longer then the forelimbs, raising the hindquarters in a characteristic sloping back appearance, and the paws are densely covered with fur in the winter to facilitate moving on snow. The head is the most heavily built of all lynx species. Each ear is darkly furred on the back with a pale central spot and has a conspicuous long, black tuft. The fur is soft and dense, with variable background colour including shades of silver-grey, yellowish, tawny and reddish-brown. The amount of spotting varies widely, falling broadly into four categories: largely unspotted in which only the lower legs and underparts are distinctly marked; small discrete dab-like spots covering the body, becoming larger on the lower limbs; large discrete spots and blotches all over the body; and elongated rosettes with dark coloured centres. There is intergradation between types and all types can occur in the same population. The winter coat in most populations is paler and less noticeably marked (as well as being longer and thicker) than in summer, and northern populations tend to be paler.

Similar species The closely related Canada Lynx is

This Eurasian Lynx in summer coat shows the most strongly marked of the four basic fur patterns that occur in the species, an elongated rosette pattern. This individual from Finland is relatively pale with faint rosettes (C).

very similar in appearance but the two species are not sympatric. The Caracal is superficially similar but is smaller, uniformly coloured and lacks distinct features of the Eurasian Lynx such as a black-tipped tail and prominent facial ruff. The two species are sympatric only in a very narrow band across south-east Turkey, north-east Iran and south-west Tajikistan.

Distribution and habitat

The Eurasian Lynx has an extensive and largely contiguous distribution in the broad temperate forest belt covering most of Fennoscandia through Russia (which contains 75 per cent of the range) to the Bering Sea coast in the east, and south into neighbouring countries from north-east China through northern Mongolia and northern Kazakhstan to Poland in the west. The distribution runs south from the Russian Altai Mountains through most of Central China (circumventing the Taklimakan Desert) and neighbouring countries along the Tien Shan Mountains and Himalayas. The distribution is patchy in northern Iran, the Caucasus and Turkey. Eurasian Lynx were extirpated from most of western Europe, with populations surviving in the Carpathian Mountains and the south Dinaric Mountains in Greece, Macedonia and Albania. Populations have been re-established by reintroduction in Austria, Czech Republic, Italy, France, Germany, Slovenia and Switzerland. Eurasian Lynx have a fairly broad habitat tolerance and occur in all types of temperate forest, open woodlands, scrub and tundra provided there is cover. They inhabit rocky habitats with sparse vegetation in montane areas (for example, in the Himalayas) and cold, rocky semi-desert (for example, on the Tibetan Plateau). They avoid very open habitats and cannot inhabit heavily modified landscapes, such as extensive agriculture, though they inhabit rural and peri-urban mosaics of forest, plantations, meadows and fields where suitable prey occurs, for example in western Europe. They occur from sea level to 4,700m in the Himalayas and exceptionally to 5,500m.

Feeding ecology

The Eurasian Lynx is the only species in the genus that specialises in ungulates. It is capable of killing prey to the size of adult Red Deer (~220kg), though their staple diet is made up of small to mid-sized ungulates and the juveniles of larger species. The most important prey across approximately half of the range are two species of roe deer: the European Roe Deer and Siberian Roe Deer, whose combined distribution overlaps around 50 per cent of Lynx distribution. Regionally, other important ungulate prey includes Northern Chamois, Siberian Musk Deer, and juvenile Red Deer, Sika Deer, Moose, ibex, Caucasian tur and Wild Boar. Lynx in the Tibetan Plateau, China, are recorded preying on Chiru (Tibetan Antelope), Tibetan Gazelle and Blue Sheep. In south-western Finland where Roe Deer were formerly absent (but now increasing), introduced White-tailed Deer from North America are abundant and form the primary prey. Ungulates are especially important prey in winter when some species (for example, Roe Deer) aggregate at feeding sites and snow increases their vulnerability to predation; exceptional kills of large ungulates including adult Sika Deer and Red Deer typically occur in deep snow

The Eurasian Lynx's broad, heavily furred feet are particularly advantageous for hunting ungulates in winter. Acting like snowshoes, the paws help keep the lynx atop a thin surface crust of frozen snow which cannot support the narrow, hard hooves of deer (C).

The Eurasian Lynx is the only lynx species that specialises on ungulate prey across most of its range. This adult female in the Jura Mountains, Switzerland has killed a male European Roe Deer, an animal around twice her weight.

especially Black-billed Capercaillie, grouse and ptarmigan. Lynx quite often kill smaller carnivores that may or may not be eaten, especially Red Fox and incidentally Tibetan Fox, Raccoon Dog, Pine Marten, Eurasian Badger, Eurasian Otter, American Mink (introduced) and Wildcat. Cannibalism is recorded rarely. Incidental prey includes amphibians, fish and invertebrates. Eurasian Lynx kill livestock, poultry and domestic carnivores. Semi-domesticated Reindeer are the primary prey in parts of northern Scandinavia where other ungulates are absent or very scarce; for example, 93 per cent of biomass consumed by Lynx adjacent to Sarek National Park, northern Sweden. In southern Norway, Lynx take significant numbers of unattended sheep, almost all of them lambs preyed upon in summer when sheep are free-ranging and Roe Deer are less available. Roe Deer, however, are the most important prey in winter. Lynx in Russia are recorded taking domestic dogs (rarely) and cats from villages; domestic cats are regularly recorded in the diet elsewhere including Finland and are sometimes killed but rarely eaten in Switzerland.

Eurasian Lynx hunt primarily at night with peaks in activity especially at dusk and less so at dawn. Diurnal activity is more common during winter and the breeding season, and when females have young kittens. Hunting is largely terrestrial; there is a record from Russia of a Lynx apparently leaping from a tree onto a female Sika Deer as it passed below. Like the closely related Canada Lynx, Eurasian Lynx search for prey by a combination of following game trails and lying in 'hunting beds' near potential ambush sites, such as along trails or at the edges of open feeding areas of ungulates. Lynx in the Bohemian Forest, Czech Republic, were found to regularly use tourist trails and their kills tended to be close to trails. Ungulates are usually killed by asphyxiation with a throat bite. There are numerous records of Lynx leaping onto the back of large ungulates and being carried up to 80m before bringing the animal down. Small prey are typically killed by a bite to the skull or nape.

Hunting success (revealed mainly by snow

with a crust that supports the Eurasian Lynx but not the deer. The availability of suitable ungulate prey declines where northern boreal forest (taiga) thins at the northern parts of the Lynx's range, especially in northern Russia and northern Fennoscandia. In these regions, lagomorphs replace ungulates as the primary prey, and the large hares, Mountain and Brown Hares, are most important. Hares, particularly the Woolly Hare, also appear to be important prey for Lynx in the Tibetan Plateau where Roe Deer or analogous species do not occur.

Throughout much of the Lynx's range, the diet becomes more diverse from spring to early autumn and small prey is more often consumed, mainly pikas, hares, small rodents, squirrels, marmots and birds, tracking) is generally high although estimates are biased towards winter when prey, especially ungulates, are more vulnerable; for example, Lynx in Sweden had very high success rates of 74 per cent hunting Reindeer, 52 per cent hunting Roe Deer and 40 per cent hunting hares. A later study in northern Sweden (where Roe Deer are absent) estimated similar rates: 83 per cent success while hunting Reindeer and 53 per cent for small species (hares, gamebirds and Red Foxes). Elsewhere in the range, success rates have been estimated at 18–43 per cent while hunting hares. On average, each individual Lynx annually kills an estimated 43 ungulates (for subadult Lynx) to 73–92 ungulates (for adult females and males). Adult Lynx preying mainly on hares in Finland each kill an estimated 120–130 a year. Eurasian Lynx cache prey by covering with snow, grass or leaves, and feed on large kills for five to seven days. They rarely scavenge, typically only when food-stressed such as during harsh winters or when debilitated. A Lynx in Russia scavenged from the carcass of a dead domestic dog. Lynx occasionally abandon carcasses to competitors including Grey Wolves, Wolverines and Wild Boar. Humans (mainly hunters) fairly regularly steal Lynx kills in Norway and Slovenia for their own use.

Social and spatial behaviour

The Eurasian Lynx is solitary and broadly territorial, with the best available information from populations in western Europe and Fennoscandia; less information is available from Russian and Asian populations. Male ranges are larger than female ranges, and overlap between males is greater than between females. Both sexes demarcate territorial boundaries with urine-marks but home ranges are generally too large to permit a high degree of exclusivity, except in small core areas. Little is known about territorial defence but encounters between adults are occasionally fatal; a resident five-year-old male in Norway was found fatally injured by another male. Range size increases from south to north reflecting the availability of prey, with the

largest ranges occurring in areas where Lynx depend mainly on hares. These populations are also subject to dramatic, cyclical fluctuations in hare numbers (as for the Canada Lynx), which may make home ranges less stable. Although not as well studied as for the Canada Lynx, hare shortages appear to provoke a similar pattern of Eurasian Lynx expanding the size of their range and abandoning it entirely if hare populations remain low for a protracted period. Presumably, as for Canada Lynx, the recovery of hare numbers permits these Eurasian Lynx to re-establish stable ranges.

The size of home ranges across the distribution varies from 98km² to 1,850km² for females and 180km² to 3,000km² for males. Range size estimates

Cats often partially pluck their prey of feathers or fur before eating, though large quantities are ingested in the course of normal feeding. It provides the cat with useful roughage and, as feathers and hair pass largely undigested, furnishes biologists with information on felid diet in the scats (C).

Three-week-old kittens in a den in the Swiss Alps. Female Lynx in Switzerland typically use about three dens for every litter, the birth den for around three weeks, followed by one to two secondary dens which kittens are carried to by the mother. By the age of eight weeks, kittens are old enough to accompany the mother and denning behaviour stops.

are best known for well-studied populations preying mainly on ungulates in which range size differences can be explained mostly by declining ungulate density from south to north. Among such populations, average range sizes for females and males respectively are: 106–168km² and 159–264km² (north-west Alps and Jura Mountains, France and Switzerland); 133km² and 248km² (Bialowieza Primeval Forest, Poland); 177km² and 200km² (Kočevje, southern Slovenia); 409km² and 709km² (Sarek, northern Sweden); 350km² and 812km² (Akershus, southern Norway); 561km² and 1,515km² (Nord-Trøndelag, central Norway); and 832km² and 1,456km² (Hedmark, southern Norway). Range size is expected to be very large in most of Russia and the Tibetan Plateau. Density estimates include 0.25 Lynx per 100km² (southern Norway), 0.4 Lynx per 100km² (Bavarian National Park, Germany), 1.5 Lynx per 100km² (Swiss Alps) and 1.9–3.2 per 100km² (Poland).

Reproduction and demography

Eurasian Lynx are seasonal breeders. Mating occurs in February to mid-April with a peak in late March. Oestrus lasts three to five days and gestation lasts 67–74 days. Births occur in May to early July. Litter size is typically one to four kittens, typically two and very exceptionally five. In captivity, young Eurasian Lynx kittens (five to nine weeks old) sometimes fight seriously, in some cases with severe injury or death of one kitten, despite the mother intervening. The reasons for this unusual behaviour are unclear and it is suspected but not confirmed for wild litters. Independence occurs at 9–11 months (rarely as young as six months) before the mother's next litter, in January to May of the following year. Most families break up in March and April at the peak of the mating period. Subadults often linger in the natal range for a few months before dispersing which usually occurs by 16 months old. Both sexes may disperse, though females are more likely than males to establish ranges within or close to those of their mothers. Information on dispersal distances is limited: 7.4–97.3km (Jura Mountains and north-eastern Swiss Alps, Switzerland) and 5–129km (Białowieza Primeval Forest, Poland). The best data come from Scandinavia where large numbers of dispersers have been radio-tracked at four sites, showing that males disperse two to five times as far as females: 15–69km (mean, with a maximum of 215km) for females, and 83–205km (mean, with a maximum of 428km) for males. Females are sexually mature at 8–12 months but the earliest that wild females breed is after their second winter at 22–24 months. In the Swiss Alps, around 50 per cent of females first breed at this age, and 50 per cent first breed a year later. Males are sexually mature at 19–24 months and will generally not breed in the wild until 33–36 months old.

Mortality Natural mortality of adult Lynx appears to be low. In Norway and Sweden, natural mortality of adults across five sites was only 2 per cent annually but this increased to 17 per cent when anthropogenic factors were included. Between 44 and 60 per cent of subadults die during dispersal (Switzerland). Annual kitten mortality is usually at least 50 per cent; 59–60 per cent of Swiss kittens die before independence.

Mortality in well-monitored populations is largely due to people, for example 70 per cent of 124 Lynx deaths in Switzerland for 1974–2002 (including poaching and accidental causes such as roadkills). Illegal killing (poaching) alone accounted for 46 per cent of known Lynx deaths from five sites in Scandinavia. Predation occurs occasionally by Wolverines (on young animals), Grey Wolves and Tigers. Domestic dogs rarely kill kittens and young animals in poor condition. Infectious disease is uncommon. The most common cause of death (aside from anthropogenic killing) in a sample of 146 dead Lynx from Sweden was sarcoptic mange, a sometimes debilitating skin condition resulting from mite infestation in which death is usually by secondary infection. Sarcoptic mange is the most common disease seen in European populations although it does not appear to have population-level impacts.

Lifespan Up to 18 years (females) and 20 (males) in the wild; up to 25 years in captivity.

STATUS AND THREATS

Globally, the Eurasian Lynx is considered secure with large areas of its massive range still relatively intact and fairly well protected or uninhabited. This is especially so in Russia where the population is estimated, roughly, to be 30,000–40,000. Mongolia and especially China also hold vast, essentially continuous areas of range but densities are lower than in forested habitat further north, and status in these countries is poorly known. Excluding Russia, the total population in Europe is estimated at 8,000, and their range has expanded significantly since the 1950s to 1960s; strongholds are in Fennoscandia (~2,800), the Carpathians (~2,800) and the Baltic States (~2,000). Populations in western Europe have undergone recovery resulting from reintroduction but all are small, isolated and considered Endangered. The relict population of around 80 Lynx in the Balkans is Critically Endangered and possibly stable due to concerted conservation efforts in Albania. Status is poorly known in central Asia.

Lynx declines are driven mainly by overhunting of their ungulate prey by people and the loss of habitat. Lynx are relatively adaptable and tolerate some human presence but they disappear from areas lacking suitable cover, prey and tolerance from people. Illegal killing by deer hunters and pastoralists keeping sheep and semi-domestic Reindeer (Scandinavia) is regarded as the main threat to the small populations in Europe. Road and rail accidents can be significant sources of mortality to populations in developed areas, for example 34 cases (of 143) in Sweden from 1987 to 2001, and 16 cases (of 124) in Switzerland from 1974 to 2002. International trade in Lynx fur was formerly significant but is technically illegal now with the exception of Russia, which sells about 1,000 furs a year. Although trade in Lynx is banned in China, Lynx fur is commonly sold domestically, and similar illegal killing for fur happens elsewhere in the Asiatic range. Sport-hunting (sometimes for fur) is legal in much of their Russo-European range; the largest numbers are taken by Russia, Estonia, Finland, Latvia, Norway and Sweden.

CITES Appendix II. Red List: Least Concern. Population trend: Stable.

11–13.9cm

● **IUCN RED LIST (2015):** Endangered
Head-body length ♀ 68.2–75.4cm,
♂ 68.2–82cm
Tail 12.5–16cm
Weight ♀ 8.7–10.0kg, ♂ 7.0–15.9kg

Iberian Lynx

Lynx pardinus (Temminck, 1827)

Spanish Lynx, Pardel Lynx

Taxonomy and phylogeny

The Iberian Lynx is classified in the *Lynx* lineage.
Some authorities formerly regarded the Iberian Lynx
as a subspecies of the Eurasian Lynx, but molecular
and morphological analysis confirms they are two
closely related and distinct species with a common
ancestor estimated about one million years ago.
There are no described subspecies.

Description

The Iberian Lynx is a long-legged, medium-sized cat similar in size to Canada Lynx and Bobcat and around half the size of the Eurasian Lynx. The head is relatively small but appears much larger due to the very prominent facial ruff that can grow to 10–12cm. The ruff is present in both sexes, with a black fringe and crisp white under the chin. The ears have black backs with a light greyish patch and long black tufts. The fur is short and coarse, and only moderately thicker in winter (in the present range). The Iberian Lynx is tawny-grey to reddish-brown with creamy buff underparts and is the most heavily spotted lynx species – all individuals are distinctly spotted, in contrast to other lynx species. The degree of spotting, however, is highly variable, from large, bold blotches and spots with conspicuous heavy lines on the neck, to small spots and dabs, which appear as uniform freckling in some individuals. There is considerable variation in spotting within populations, and a wide range of intermediate forms between the two extremes. The tail has black fur completely encircling the tip.

Similar species The Iberian Lynx is very similar to spotted forms of the Eurasian Lynx but is about half the weight and the two species are not sympatric; the nearest Eurasian Lynx population is approximately 1,200km away in the Jura Mountains on the France-Switzerland border.

Distribution and habitat

The Iberian Lynx is restricted to two breeding populations 150km apart in Andalucía, southern Spain. The larger population lives in a core area of 260km² in the Sierra Morena mountains. Two associated small satellite populations have been re-established by reintroduction approximately 40–50km east and west of the main Sierra Morena population. Females are breeding in the satellites and individuals are moving between them and the main population, significantly expanding the total area used by the cats. The second main population lives in approximately 443km² in the Doñana National Park and surrounding areas. The Iberian Lynx has not occurred in Portugal since the early 1990s, although reintroduction is underway, with the first individuals released in 2015.

Iberian Lynx are restricted to mosaics of Mediterranean scrubland and forests of oak and olive trees, with interspersed open meadows. Lynx favour

the edges between closed habitat and open areas for hunting although they avoid extensive tracts of open habitat. The Sierra Morena population prefers areas rich in rocky granite outcrops that provide cavities for refuges and breeding dens, and are associated with higher rabbit numbers. Iberian Lynx avoid landscapes lacking a certain amount of understorey and/or sufficient rabbit numbers including many

Compared to many felids, the Iberian Lynx's facial regalia are dramatic. Flattening the ears and revealing the pale back-patch enhances the visual signal created by the cheek ruffs and tasselled ears.

anthropogenic landscapes such as *dehesas* (human-made open savannas cleared of scrub to promote grazing for cattle and wild ungulates), extensive agriculture and open plantations. However, they do well in some modified landscapes that give rise to high rabbit numbers. They use exotic plantations for dispersal and they are able to permanently inhabit plantations provided understorey is maintained and rabbit densities are high. They are recorded breeding in olive tree plantations with patches of scrubland.

Feeding ecology

The Iberian Lynx is greatly reliant on the European Rabbit, which comprises 75–93 per cent of the diet, and they cannot live permanently in habitats that lack abundant rabbits. Lynx prey almost exclusively on rabbits regardless of their abundance; for example, rabbits comprise around 90 per cent of the

diet by volume in two areas of Sierra Morena, despite the western area having three times as few rabbits as the eastern area. Lynx target juvenile rabbits when they are available, probably because they are easier than adults to catch; 75 per cent of rabbits killed during May–June are juveniles (Sierra Morena). Incidental prey includes birds, small rodents, hares and reptiles. After rabbits, the most important prey is the Red-legged Partridge in Sierra Morena, and ducks and geese (mainly Mallard) in Doñana. Fallow and Red Deer are sometimes killed, generally during autumn and winter. Juvenile deer are usually killed, though there are records of adult Fallow Deer (of both sexes) and one record of a Red Deer doe killed by a Lynx. Iberian Lynx frequently kill other carnivores including Red Foxes, Egyptian Mongooses, Common Genets and feral domestic cats but these are rarely eaten. They are presumably

Female Iberian Lynx in the Doñana National Park nearly always give birth to their litters inside confined, natural hollows in Cork Oaks and Narrow-leafed Ash trees. Kittens usually stay in the birth den for 20 to 36 days before being moved to secondary den sites, probably related to the growing kittens' need for more space as they develop their motor skills.

killed chiefly as competitors for rabbit prey; numbers of mongooses and genets in areas of high Lynx density are 10–20 times lower than in areas where Lynx are absent. Iberian Lynx kill domestic poultry and lambs when available.

Hunting by Iberian Lynx closely follows the activity patterns of rabbits and thus is mainly nocturnal with crepuscular peaks. Diurnal activity increases when temperatures are cool, and during overcast or rainy weather. Lynx search for rabbits in productive areas such as the transitional habitat between dense areas to open ones, along roads and firebreaks, and near warrens. Rabbits are killed by biting the skull, while larger prey such as young deer are killed by asphyxiation with a throat bite. Hunting success rates are unknown although Lynx kill a rabbit every 1–1.5 days. Adult Lynx require the equivalent of 277 rabbits per year for a female

without kittens to 379 rabbits for a male. Lynx drag large kills such as deer carcasses into dense scrub and may cover them with leaf-litter and soil debris to consume over a number of days. Iberian Lynx occasionally scavenge and they abandon kills if disturbed by competitors, primarily Wild Boar.

Social and spatial behaviour

The Iberian Lynx is mostly solitary and territorial. They establish small, stable ranges with exclusive core areas and intrasexual overlap at the edges. Male ranges are slightly larger than those of females, and each male range overlaps all or part of one to three female ranges. Territorial fights between Lynx, usually a resident and an immigrant searching for territory, are violent and occasionally fatal. Lynx occasionally acquire territory when two years old but high-quality territories are typically acquired at three to seven (females) and four to seven years (males). As occasionally observed in other Lynx species, adult female Iberian Lynx sometimes maintain amicable contact with their grown offspring (mainly with daughters and occasionally with sons), including observations of sharing kills; for example, a resident adult female associated with her grown adult daughter and independent subadult son from a subsequent litter; all three animals occasionally fed on the same deer kills. There are frequent observations of a mother Lynx sharing her home range with a grown daughter, and both bringing their litters together for extended periods in which only one of the females accompanies the kittens.

Home range size is 8.5–24.6km² averaging 12.6km² for females, and 8.5–25.0km² averaging 16.9km² for males (Doñana). Density estimates range from 10–20 adults per 100km² in habitat with moderate rabbit densities to exceptionally high densities of 72–88 Lynx per 100km² in optimal habitat. Importantly, such high densities only occur in very small enclaves with excellent habitat and very high rabbit densities, for example in the 8km² 'Coto del Rey' area of the Doñana National Park.

The genus *Lynx* has the most specialized prey requirements of all felid genera. This is especially true of the Iberian Lynx which cannot survive without ample populations of the European Rabbit.

Reproduction and demography

The Iberian Lynx breeds seasonally, which is somewhat unexpected given the mild seasonal extremes in its range (compared to other Lynx species). This may be an evolutionary artefact given the predominance of seasonal reproduction in the *Lynx* genus, but is perhaps also influenced by strong seasonal breeding patterns of the European Rabbit. Iberian Lynx mate chiefly in December to February with births peaking in March, and occasional births as late as July. Gestation lasts 63–66 days. Litter size is two to four kittens, averaging three. Like Eurasian Lynx, captive young Iberian Lynx kittens (6–11 weeks old) sometimes engage in serious fights that occasionally end in serious injury or death to one kitten, despite the mother intervening. The reasons for this unusual behaviour are unclear and it is suspected but not confirmed for wild litters. Kittens reach independence at seven to eight months. They linger in their natal range until dispersal, which rarely occurs before one year of age and typically at 13–24 months (average 17.8 months). Dispersal occurs mainly in the first half of the year, apparently coinciding with increased social activity of residents during the breeding season. Dispersing Lynx in the Doñana ecosystem settle a straight-line distance of 2–64km away from the natal range, after walking an average of 172km. Around 50 per cent of dispersers (whose fate is known) successfully disperse and settle in a new territory. Both sexes can breed at two years but reproduction in the wild is usually at age three or later, associated with establishing a territory. Females breed until age nine in the wild.

Mortality Annual adult mortality was formerly estimated at 37 per cent in Doñana (1983–1989) but this has been considerably reduced with intensive

Except for Critically Endangered Grey Wolves in the Sierra Morena population, the Iberian Lynx is exposed to few natural predators. Small kittens might be vulnerable to large eagles and owls, or to mesopredators such as Wildcats and Red Foxes.

conservation effort. Annual mortality (subadults and adults combined) is now approximately 12 per cent in Doñana and 19 per cent in Sierra Morena (16 per cent for the entire population combined; 2006–2011). Subadults have higher annual mortality rates (24 per cent) than adults (14 per cent). Kitten mortality is around 33 per cent; two kittens from most litters of three usually survive to independence. The mortality rate escalates during dispersal. During the 1990s, comparing Doñana Lynxes of the same age that stayed in their well-protected, natal range versus those which dispersed, mortality increased from 8 per cent to 52 per cent for Lynx aged 12–24 months; the same pattern exists today although the mortality rate for dispersers is not as high. Historically, humans were the main cause of mortality; this is still true for the Sierra Morena population, chiefly from incidental trapping (in snares or traps set for rabbits and foxes), intentional shooting and road accidents. Infectious disease is the main mortality factor for the Doñana population, possibly due to relatively high levels of inbreeding (which can elevate the vulnerability to disease) as well as high population densities (which facilitate transmission). Deaths from at least six different pathogens are recorded in this population, including a lethal outbreak of feline leukaemia virus in 2007 (transmitted presumably by infected domestic cats) in which seven Lynx died before rapid intervention, including vaccination and the removal of infected animals and cats. Iberian Lynx have no natural predators in their remaining range. They are occasionally killed by domestic dogs (usually hunting dogs accompanying poachers).

Lifespan Up to 13 years in the wild and up to 20 years in captivity.

STATUS AND THREATS

The Iberian Lynx is the world's most endangered felid, at least in terms of numbers. It formerly inhabited the entire Iberian Peninsula, bounded in the north by the Pyrenees. By the 1940s, the species occurred only in southern Spain and parts of Portugal, and had been reduced to its present two populations in Andalucía by the 1990s. By 2002, there were between 84 and 143 adults in these two populations occupying 2 per cent of its historical range. The decline has been driven by massive habitat conversion of natural forest and Mediterranean scrublands for human use, chiefly agriculture and exotic plantations. This has been compounded by the introduction of myxomatosis into rabbits in Europe, which resulted in a massive crash of the Lynx's principal prey. This was followed more recently by outbreaks of viral haemorrhagic pneumonia in rabbit populations, which has kept their numbers very low. Finally, anthropogenic killing is a serious threat. Poaching was the primary cause of extinction in many areas of the cat's former range and humans are still responsible for many Lynx deaths, mostly by illegal trapping, shooting and roadkills.

The Iberian Lynx has been the focus of targeted conservation efforts since 1994 and was declared Critically Endangered in 2002. In the same year, a massive, renewed effort to save the Lynx was launched, entailing programmes that reduced human killing of Lynxes and improved habitat quality and rabbit populations, and a captive-breeding programme for reintroduction to the wild. Although the situation of the Lynx remains perilous, these actions have successfully reversed the decline and significantly improved the species' status. By 2010, the minimum number of known Lynx in the wild increased to 252 and the area occupied by breeding Lynx increased from 293km^2 to 703km^2. Anthropogenic mortality has been targeted by increasing enforcement of protected areas, building underpasses for Lynx to avoid major roads, and education programmes that build tolerance among private landowners where 80 per cent of Lynx live. Human-caused mortality decreased from 40 per cent (1992–1995) to 7.4 per cent (2006–2010) in Sierra Morena; and from 58.4 per cent (1983–1989) to 11.1 per cent (2006–2010) in Doñana. Despite these successes, the threats of ongoing habitat loss, anthropogenic mortality, disease outbreaks and low rabbit numbers remain.

CITES Appendix I. Red List: Endangered. Current population trend: Increasing.

10.6–13.7cm

IUCN RED LIST (2008):
Least Concern

Head-body length ♀ 50.8–95.2cm,
♂ 60.3–105cm
Tail 9–19.8cm
Weight ♀ 3.6–15.7kg, ♂ 4.5–18.3kg

Bobcat

Lynx rufus (Schreber, 1777)

Bay Lynx

Taxonomy and phylogeny

The Bobcat is classified in the *Lynx* lineage, where it is believed to be the earliest species to diverge and is more distantly related to the other three *Lynx* species than they are to each other. Bobcats are known to hybridise with Canada Lynx where the two species overlap. Hybrids have been documented from Maine, Minnesota and New Brunswick, all of them resulting from a male Bobcat mating with a female Lynx. Hybrids were formerly considered to be sterile but at least two female hybrids are known to have reproduced. Twelve Bobcat subspecies are currently described based largely on superficial differences that are unlikely to be valid. Genetic analyses of populations in the contiguous US suggest two subspecies, eastern and western, with a transition zone in the Great Plains grasslands of the central US.

Description

The Bobcat is a medium-sized, stocky felid around two to three times the size of a domestic cat. The body is relatively robust with long legs and a short tail that rarely exceeds 20cm. The head is relatively small with a prominent facial ruff and triangular, black-backed ears with a white spot and short black tufts. The ear-tufts, however, are quite often inconspicuous or missing entirely. Body size varies broadly along a cline, increasing with latitude and elevation so that the largest, heaviest individuals occur at the northern limit of the range, for example in Minnesota, British Columbia and Nova Scotia. Sexual dimorphism is pronounced in the Bobcat, with males weighing 25–80 per cent more than females in the same population. The Bobcat's fur is short, soft and dense with a variable background colour of various shades of frosted grey to rich rust-brown, with markings ranging from faint freckling to large Ocelot-like blotches. Northern Bobcats tend to be paler with minimal markings, especially in the long winter coat. The Bobcat is the only *Lynx* species in which melanism occurs, mostly recorded from Florida, and albino individuals occur very rarely.

Similar species The Bobcat and Canada Lynx are easily confused and co-occur along a broad swathe

Bobcats can be richly spotted with large blotches as in this female and her four-month-old kittens in California. Contrary to popular myth, this is not the result of hybridisation with the Ocelot.

The relatively small face and gracile upper body of this Bobcat are typical of adult females. Males are more heavily built with broader, more muscular heads.

the densely populated North-east and from the intensively cultivated Midwest, but has recolonised much of this former range and it now has a presence in every US state except for Delaware. It has re-established populations that are increasing in the Midwestern states where it was previously extirpated (Iowa, Illinois, Indiana, Missouri and Ohio). In the 20th century, the Bobcat expanded its range into northern Minnesota, Ontario, New Brunswick and Manitoba as a result of people converting boreal forest into open habitat. It currently inhabits all southern Canadian provinces.

The Bobcat has an extremely wide habitat tolerance and occurs in virtually any habitat provided there is cover in the form of thick vegetation or broken terrain. It inhabits all types of forest, grassland, prairie, brushland, scrub, semi-desert, desert, marshland, swamp, coastal habitat and rocky habitat. They avoid areas with very deep snow and the northern extent of their range is delimited by heavy snow accumulation. They occur up to 2,575m in the Rockies (US) and to 3,500m on the Colima Volcano in Colima/Jalisco, Mexico. Bobcats readily inhabit anthropogenic habitats including many kinds of farmland but they avoid extensive stands of open agriculture or pasture. Bobcats are able to live close to people including in urban landscapes provided there is cover such as parkland or intact riverine habitat.

Feeding ecology

The Bobcat is a powerful and opportunistic hunter, recorded killing adult White-tailed Deer weighing up to 68kg, but the majority of the diet is made up of small vertebrates weighing 0.7–5.5kg. Throughout most of the range, Bobcat diet is dominated by lagomorphs especially Snowshoe Hares, cottontails and jackrabbits, which comprise up to 90 per cent of the diet. Depending on the region and season, the amount of lagomorphs in the diet is supplemented or exceeded by other prey types, especially deer or rodents. Bobcats readily prey on White-tailed Deer, Mule Deer, Pronghorn and Bighorn Sheep, primarily

of the Canada-US border and south through the Rocky Mountains. The Lynx is generally larger and taller but the largest Bobcats occur in the areas of sympatry; for example, male Bobcats on Cape Breton Island, Nova Scotia, outweigh male Lynx by an average of 40 per cent. The Bobcat's tail has three to six dark half rings with a vivid white underside and tip, compared to the Lynx's unstriped tail with a solid black tip.

Distribution and habitat

The Bobcat occurs in an almost continuous range across southern Canada, throughout the contiguous US and throughout Mexico to Oaxaca state. Bobcat range is thought to stop at the Isthmus of Tehuantepec in southern Mexico based on habitat modelling. The species was extirpated in the US from as fawns but kills of healthy adults are not uncommon. Male Bobcats kill more adult deer than do females; many Bobcat populations show a marked sex-based difference in the diet in which both sexes eat mainly lagomorphs, which females supplement with smaller prey such as rodents, and males supplement with larger prey such as deer.

Adult deer are most often killed in the northern range and at higher elevations where Bobcats are large and cold winters provide an advantage. Consumption of deer peaks in the winter when they are most vulnerable due to deep snow and nutritional stress. Outside winter and generally in the rest of the range, fewer adult deer are killed, and fawns often comprise an important part of the diet. For example, Bobcats killed 23 per cent of Pronghorns born during a five-year period in western Utah, and Bobcats were responsible for 67 per cent of White-tailed Deer fawn deaths during a study on Kiawah Island, South Carolina. Bobcats eat a wide variety of rodents. Cotton rats (genus *Sigmodon*), woodrats (genus *Neotoma*) and kangaroo rats (genus *Dipodomys*) are important prey in the south-eastern and south-western US, while North American Porcupine and Mountain Beaver are important in New England and Minnesota, and Washington state, respectively. Squirrels, including Woodchuck, as well as North American Beaver and Muskrat are also eaten. Bobcats living in an urban-rural landscape in northern California make up one of the few populations to eat mainly small rodents, in this case California Voles, which are abundant. Bobcats kill a variety of birds and reptiles, though the incidence of both in the diet is unexpectedly low. The number of reptiles in the diet increases from north to south, reaching a maximum of around 15 per cent of intake in the south-eastern US. Incidental prey of Bobcats includes bats, Virginia Opossum, smaller carnivores (for example Kit Fox, White-nosed Coati and various mustelids), juvenile Javelina and feral pigs, amphibians, fish, arthropods and eggs. Cannibalism is recorded rarely. Bobcats kill sheep, goats, piglets and poultry, though they rarely create major damage

except in particular circumstances such as in untended lambing areas where large numbers of young stock may be killed. Bobcats were recorded preying on young Japanese Macaques in a fenced, semi-wild research colony in Texas. Small pets are rarely attacked in suburban and rural areas, and apparently not as prey; domestic cats do not appear in the diet.

Bobcats forage mainly at night with crepuscular activity peaks, but this varies widely depending on the region and season. They tend to be more diurnal during winter and in areas where they are not persecuted; some populations are cathemeral and foraging may occur at any time. Most hunting takes place on the ground but the Bobcat is a very capable climber that readily pursues arboreal prey such as squirrels taking refuge up trees. They also hunt in shallow water for fish or amphibians, and launch attacks at waterfowl on the water. Like many felids, Bobcats hunt primarily by two techniques: moving steadily through their territory searching for prey by sight or sounds and waiting in ambush at strategic sites such as burrow entrances, along game trails, on ledges and near water sources. Most hunts are preceded by a careful stalk culminating in an

A young Bobcat hunting rodents in grassy habitat. Habitat edges – where one vegetation type transitions to another – including along trails or roads provide fine hunting opportunities for foraging Bobcats.

The stage in a cat's life when littermates are maturing together will be the most intensely social period for individuals of most species. It is a period in which constant play lays the foundations for important survival skills.

explosive rush at the quarry, typically from within 10m. While hunting rodents in long grass, Bobcats move slowly into position after locating the prey by sound and attempt capture by a high, arching pounce. Small prey species are killed by biting the skull or nape, while larger prey, such as ungulates, beavers and porcupines, are usually asphyxiated with a throat bite.

For such a well-studied species, Bobcat hunting success rates are poorly known. They sometimes cache carcasses with a covering of dirt or snow to consume over time, for example up to 14 days for an adult deer kill during winter. Bobcats eat carrion; road- and winter-killed deer may be an important food source in northern winters.

Social and spatial behaviour

The Bobcat is essentially solitary and territorial. Adults socialise mainly when mating but resident males interact amicably with familiar females and kittens. Both sexes establish enduring ranges with exclusive core areas and considerable overlap at the edges. The ranges of males are typically two to three times the size of female ranges but occasionally up to

five times as large, for example in Oregon and Maine. Territories are demarcated mainly by urine-marking and scrapes, and defended from same-sex conspecifics; fights are rare but occasionally fatal. The Bobcat's extensive distribution and very wide habitat tolerance produces a similarly wide variety in spatial characteristics and population density. In general, ranges are largest, overlap is greatest and density is least in northern latitudes, probably due to low prey density and larger Bobcat body size with concomitant energetic demands. Well-protected populations with mild climates in very productive habitat have the smallest ranges and highest densities, for example in California and Kiawah Island, South Carolina.

Average range size varies from 1–2km^2 (Alabama, California, Louisiana) to 86km^2 (Adirondack Mountains, New York) for females, and 2–11km^2 (Alabama, California, Louisiana) to 325km^2 (Adirondack Mountains, New York) for males. Range size contracts during peaks in the densities of prey, especially of hares and rabbits. Nightly movements can be as large as 20km but are typically 1–5km. Density estimates include 4–6.2 bobcats per 100km^2

(Idaho, Minnesota, Utah), 6–10 per 100km² (Missouri), 20–28 per 100km² (Arizona, Nevada) and exceptionally exceeding 100 Bobcats per 100km² in areas with both high prey availability and protection from hunting (coastal California and Kiawah Island, South Carolina).

Reproduction and demography

Bobcats are able to reproduce year-round although most mating occurs from December to July with births peaking in spring or summer, strongly so in northern areas. Females that breed early are able to have a second litter in the same year, generally restricted to southern populations where conditions are mild. Oestrus lasts 5–10 days, gestation is 62–70 days. Litter size averages two to three, and exceptionally reaches six. Weaning occurs at two to three months. Kittens reach independence at 8–10 months, and disperse at 9–24 months. Dispersing Bobcats in Montana travelled an average of 6.6km before dying or settling compared to an average distance of 33.4km in Missouri. The longest documented dispersals were by two young males in Idaho that travelled 158km and 182km following a crash in the jackrabbit population. Female Bobcats can breed at 9–12 months but usually first give birth after 24 months. Males are able to reproduce by their second winter (approximately 12–18 months) although few sire kittens until they become territorial residents from around age three onwards.

Mortality Annual adult mortality varies from 20 to 33 per cent for unharvested populations to 33 to 81 per cent for hunted populations. Kitten mortality fluctuates extensively, depending chiefly on prey numbers (and thus risk of starvation) and level of human hunting; for example, 29–82 per cent of kittens died each year in Wyoming in different years of monitoring. Bobcats are mostly killed by people; legal hunting as well as illegal and accidental killing are the primary mortality factors for most populations. The most important natural causes of death are winter starvation and predation by other carnivores, mainly Pumas and Coyotes (as well as domestic dogs in rural/urban populations). There are also rare records of predation by Grey Wolf, Golden Eagle (the latter of kittens), American Alligator and introduced Burmese Python. Episodic outbreaks of infectious disease in dense populations occasionally produce significant mortality; feline panleukopenia was responsible for 17 per cent of mortalities in a high-density Californian population. An epizootic of notoedric mange (feline scabies, arising from mite infection) elevated the annual mortality rate of Bobcats in southern California from 23 per cent to 72 per cent in two years. This population was heavily exposed to anticoagulant rodenticides, thought to have increased its vulnerability.

Lifespan Up to 23 years in the wild (typically much less) and up to 32.2 in captivity.

STATUS AND THREATS

Bobcats are widespread, resilient to human pressures and secure in most of their range. The total population is unknown although it is likely to be greater than 1.4 million in the US alone. Based mainly on harvest data and hunter surveys (the reliability of which are variable), a 2010 analysis estimated the population in the contiguous US at 2.35–3.57 million. Nonetheless, many populations are exposed to intense hunting pressure and the species is vulnerable to overharvest when this is poorly monitored. Around 50,000 Bobcats are legally killed in the US and Canada annually, for recreation and fur. With trade restrictions on spotted cat furs, the Bobcat is now the most heavily traded felid species for its fur. World demand for its fur is at a record high since the 1960s, driven in large part by new markets in China and Russia. Fur hunting can lead to population declines when it is poorly managed and during years with harsh winters; the trapping season in North America often occurs around winter, coinciding with high rates of natural mortality especially when lagomorph numbers decline in poor years. Bobcats are also persecuted for supposed livestock depredation (for example, in Mexico) and 2,000–2,500 are killed annually in legal predator control in the US in response mostly to depredation complaints. There is anecdotal evidence that Bobcat numbers have declined in southern Florida with the spread of invasive Burmese Pythons, and exposure to anticoagulant rodenticides in southern California is believed to have mediated a population crash by increasing its vulnerability to disease.

CITES Appendix II. Red List: Least Concern. Population trend: Stable (considered to be increasing in much of the US and southern Canada).

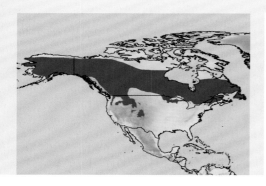

11.7–13.9cm

● **IUCN RED LIST (2008):**
Least Concern

Head-body length ♀ 76.2–96.5cm,
♂ 73.7–107cm
Tail 5–12.7cm
Weight ♀ 5.0–11.8kg, ♂ 6.3–17.3kg

Canada Lynx

Lynx canadensis (Kerr, 1792)

Canadian Lynx

Taxonomy and phylogeny

The Canada Lynx is classified in the *Lynx* lineage. It is thought to have evolved from a common ancestor with the Bobcat in a relatively recent separation that also led to the Canada Lynx's closest relative, the Eurasian Lynx. Some authorities have classified the Canada Lynx and Eurasian Lynx as the same species but molecular analysis confirms they are distinct with divergence around 1.5 million years ago. Canada Lynx rarely hybridise with Bobcats where the two species are sympatric, with hybrids recorded from Maine, Minnesota and New Brunswick. All known hybrids were born to a female Lynx with a male Bobcat sire, and are widely regarded as infertile, but at least two female hybrids are known to have reproduced. Two Canada Lynx subspecies are described: *L. c. canadensis* on mainland North America and *L. c. subsolanus*, an isolated population on Newfoundland island, though the morphological differences between them are superficial and have been disputed.

Description

The Canada Lynx is a lanky, medium-sized cat whose thick fur gives the impression of a much larger animal. On average, Canada Lynx are taller and slightly larger than Bobcats, though the largest Bobcats are heavier than the largest Lynx. The Canada Lynx has elongated legs with hindlimbs distinctly longer then the forelimbs giving a characteristic leggy appearance. This is enhanced by the very large, snowshoe-like paws which have loosely knit metatarsals enabling the toes to spread widely, and a heavy covering of dense fur, to facilitate moving on soft snow. The head is relatively small but appears larger due to a prominent facial ruff. The triangular ears are unmarked except for a band of black fur along the top edge, capped with conspicuous black tufts typically longer than 3cm. The long fur is dense, very soft and uniformly

Left: **All species of lynx have excellent vision, giving rise to a rich mythology that includes the ability to see through objects. While lynx vision is superb, there is no evidence to suggest it is any more acute than that of any other felid.**

Below: **An adult female Canada Lynx in her short, summer coat. With the onset of winter, long, densely packed, pale coloured hairs will replace her tawny-brown fur, transforming her colouration to grey or greyish-brown.**

The Canada Lynx is the only felid that undergoes cyclical population crashes, linked to cycles of growth and decline in hare numbers. Crashes occur roughly every 10 years and are most extreme for northern Canada Lynx populations which are almost entirely reliant on hares as prey.

coloured without distinct markings, typically buff-grey with silver or bluish frosting in winter and brownish in spring and summer. The lower limbs and underparts are sometimes lightly spotted. The tail is shorter than in Bobcats and has a completely black tip.

Similar species   The Canada Lynx and Bobcat are easily confused where they co-occur along the Canada-US border and through the Rocky Mountains. The Lynx is generally larger, taller and less spotted, but the largest Bobcats occur in the areas of sympatry. The Lynx's tail is uniformly coloured with a completely black tip compared to the Bobcat's tail that has three to six dark half stripes and a vivid white underside and tip.

Distribution and habitat

The Canada Lynx occurs in most of Canada south of the tree-line (comprising around 80 per cent of its range), most of Alaska (13.5 per cent of range) and within the contiguous US in the southern extensions of the boreal and sub-boreal forest along the Rocky Mountains, Cascades and Blue Mountains (Washington, Oregon, Idaho, Montana and Wyoming), Great Lakes region (Minnesota and

Wisconsin) and New England (Maine, New Hampshire and northern Vermont). They have been successfully reintroduced into Colorado where their southernmost occurrence now extends into northern New Mexico. An attempt to reintroduce them into the Adirondack Mountains, New York (1989–1992), was unsuccessful. Canada Lynx are habitat specialists and occur only in dense boreal and coniferous forests of aspen, spruce, birch, willow, fir, poplar or pine, essentially matching the distribution of the main prey species, the Snowshoe Hare. Canada Lynx are extremely well adapted for snow and ice. They are recorded swimming the Yukon River up to 3.2km wide, and two collared Lynx routinely swam for four to 12 minutes across extremely, hazardous semi-frozen rivers when air temperatures were as low as -27°C . Canada Lynx avoid open habitat, even though it often contains abundant prey. They rarely live in heavily modified habitat such as agriculture and they do poorly where forestry permanently thins out cover or reduces forest complexity. They do well in forest left to regenerate after clear-cutting or intensive logging, provided recovery has been allowed to proceed for around 15 years or more. The Canada Lynx occurs from sea level to 4,130m.

Feeding ecology

Canada Lynx are strongly dependent on Snowshoe Hares which constitute 35–97 per cent of the diet. Every Lynx population relies primarily on the hare, but the percentage in the diet fluctuates in different seasons and years, depending on hare abundance. Northern Snowshoe Hare populations cycle every 8–11 years, sometimes spectacularly from as high as 2,300 hares per km² to as low as 12 per km². Lynx numbers are closely linked to hare density, lagging one to two years behind with a three- to 17-fold drop, for example from 17 lynx per 100km² to 2.3 per 100km² (south-western Yukon, Canada). Lynx switch prey during hare declines, though hares usually remain the primary prey, for example from 97 per cent of the diet by biomass when hares are abundant to 65 per cent of the diet when they are scarce (central Alberta). Lynx diet becomes more diverse during summer and autumn (in the northern range), and is more diverse year-round for southern Lynx populations where hares occur in low to moderate densities and have weak or non-existent population cycles; Snowshoe Hares remain the primary prey regardless. Red Squirrels are particularly important as an alternative food source during hare declines, rising from 0–4 per cent of the diet to 20–44 per cent by biomass (south-western Yukon). Other common alternative prey includes small rodents, especially mice and voles, and ducks, grouse and ptarmigan. Lynx occasionally prey on juvenile ungulates, mostly during fall and winter and/or during hare declines. Ungulates usually constitute a minor part of the diet, though Lynx on Newfoundland island specialised on Caribou calves during a hare crash. Lynx in the same population were recorded attacking winter-stressed adult Caribou does and stags, in some cases leading to their deaths, though it was unclear if Lynx successfully brought down adults in a single event. Other ungulates recorded in Canada Lynx diet include deer, Moose, Bison and Dall's Sheep, though scavenging probably accounts for many records. Incidental prey records include flying squirrels, ground squirrels, North American Beaver, Muskrat, Red Fox, American Marten and fish. Cannibalism occurs rarely, primarily during prey shortages in which young transient Lynx (typically) are killed by adults apparently as prey. Canada Lynx are occasionally recorded killing small lambs or poultry.

Hunting is mainly crepuscular and nocturnal, reflecting activity patterns of Snowshoe Hares, and occurs almost exclusively on the ground. Canada Lynx are capable climbers that may seek refuge in trees when threatened, but they do not hunt arboreally. Lynx typically follow well-used hare paths or lie in wait in 'hunting beds' at strategic ambush sites, such as along trails or in cover near open patches where prey aggregates. Hares are killed by biting the skull, nape or throat while juvenile ungulates are usually asphyxiated with a throat bite. Newfoundland Caribou calves attacked by Lynx on Newfoundland island often escaped, perhaps due to the doe intervening, but most died later from *Pasteurella* bacterial infection transmitted in Lynx saliva. It is tempting to think this was a strategy to provide the Lynx with a carcass, though many dead calves were not fed on. Hunting success (revealed by snow-tracking) varies, depending mainly on the bearing strength of the snow and, in some cases, hare abundance, but rates are generally high, ranging from 24 to 61 per cent (all for hunts of hares). On average, Lynx kill one hare every one to two days. Canada Lynx cache prey by covering with snow or leaves, and this occurs more frequently when hares are abundant. They readily scavenge, especially on winter- and roadkilled ungulates. A

A young Canada Lynx takes refuge in a snow-covered tree, Yukon Wildlife Preserve, Canada. The Canada Lynx does the majority of its hunting on the ground and is not recorded hunting in trees except for opportunistically pursuing fleeing, arboreal prey like squirrels into lower branches.

females, and they overlap multiple female ranges. These differences break down in northern populations during low hare abundance.

Resident female Lynx sometimes maintain amicable contact with their grown daughters (which often settle near the mother's range; see next section), in some cases, for life. Female cats of many species often relinquish part of their range to daughters although the degree of long-term, amicable interactions in Lynx is unusual for felids. Numerous observations exist of females associating amicably with a grown daughter, often while both are accompanied by kittens. Some cases include hunting together and sharing kills.

Home range size of the Canada Lynx varies from $8km^2$ to $738km^2$. Average range size estimates include $39-133km^2$ (females) and $69-277km^2$ (males) in southern populations; $13-18km^2$ (females) and $14-44km^2$ (males) in northern populations with high hare numbers; and $63-506km^2$ (females) and $44-266km^2$ (males) in northern populations with few hares. Lynx density varies drastically depending on hare availability. Densities in the southern range tend to be low and stable, at two to three Lynx per $100km^2$, which is also typical of northern populations during low hare abundance. Density in northern populations climbs to $8-45$ Lynx per $100km^2$ during peaks in hare abundance. The highest densities, $30-45$ Lynx per $100km^2$, occur during hare peaks in regenerating northern forest (more than $15-20$ years post-logging), which provides excellent habitat for hares. Densities in mature northern forest during hare peaks are $8-20$ Lynx per $100km^2$.

Reproduction and demography

Breeding in the Canada Lynx is strongly seasonal. Most mating occurs in March to early April, though occasionally as late as May. Oestrus lasts three to five days and gestation lasts $63-70$ days. Births occur from May to early July with most litters born mid-May to early June. A female in Maine gave birth to a single kitten in August, apparently after an earlier litter failed, the only known case of a female

Climatic warming at the southern periphery of Canada Lynx range is already having an impact on the species. Since the 1970s, Lynx distribution in central Canada has receded northwards by more than 175km as warmer summers result in reduced snowfall and an opening up of habitat.

Lynx in south-eastern British Columbia fed on a roadkilled Mule Deer for four days until it was usurped by wolves.

Social and spatial behaviour

The Canada Lynx is solitary and loosely territorial but spatial behaviour varies considerably depending on hare availability. Southern populations with stable but low densities of hares tend to maintain large, enduring home ranges with high overlap between neighbours. Northern populations maintain smaller and possibly more exclusive ranges during hare peaks but ranges expand during declines, sometimes leading to complete abandonment of the range and large, nomadic movements. Males in southern populations and in northern populations during high hare abundance tend to have larger ranges than producing two litters in the same season. Lynx have large litters, with as many as eight kittens recorded in the wild. During hare peaks, female Lynx breed younger (in their first spring), more females breed successfully and larger litters are born, averaging four to five kittens compared to one to two during hare shortages. Yearling females rarely if ever breed and adult females do not breed if hare availability is very low for successive years. Independence occurs at 10–17 months and dispersal of subadults usually occurs in the spring and summer. Female dispersers sometimes establish ranges within or close to those of their mothers, and adult females may retain intermittent contact with their female offspring for life (see previous section). Male dispersers tend to move more widely and settle far from their natal range if they survive. Dispersal also occurs among resident adults during protracted hare shortages. Both subadult and adult Lynx are capable of extremely long dispersal movements, particularly during hare declines; the longest known straight-line dispersal was 1,100km (Yukon, Canada). Females are sexually mature at 10 months and may breed at this age during high hare years, but the first litter is more commonly at 22–23 months old. Males are able to reproduce by their second winter (approximately 18 months) although it is assumed males do not reproduce until their second or, more often, their third spring.

Mortality Estimates of annual adult mortality during high hare abundance include 11–30 per cent for unharvested or lightly trapped populations, which can increase to 60–91 per cent during hare shortages. Annual adult mortality for hunted populations ranges from 45 to 95 per cent. Kitten survival is linked closely to hare numbers, with 17–50 per cent mortality during high hare availability and reaching 60–95 per cent mortality in poor years when starvation results in very high losses. Starvation during winter and trapping by people are responsible for most deaths of all age groups. Known predators include Grey Wolf, Coyote, Wolverine, Cougar and rarely Bobcat. Other Lynx constituted the most common predator during a hare shortage in one study. Although disease is not a common mortality factor, seven Lynx reintroduced in Colorado died from sylvatic plague, thought to have been caught from prey.

Lifespan Up to 16 years in the wild (rarely past 10) and up to 26.9 years in captivity.

Trapping of Canada Lynx for their fur is legal in Canada and the United States (mainly Alaska) including by snaring. Around 10,000–15,000 lynx are legally trapped each year. In Canada alone, annual harvest regularly topped 40,000 lynx as recently as 1980, and was as high as 80,000 per year during historical trapping highs prior to 1900.

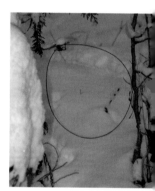

STATUS AND THREATS

The Canada Lynx is widespread and generally common in most of Canada (where it still occupies an estimated 95 per cent of its historical range) and Alaska. It has lost some range from southern Alberta, Saskatchewan and Manitoba and is uncommon to rare in eastern Canada where two provinces (New Brunswick and Nova Scotia) classify the Lynx as Endangered. It is extinct on Prince Edward Island and in mainland Nova Scotia; it still occurs on Cape Breton Island. Range loss is much more extensive in the contiguous US where it once occurred in 24 states and is now restricted to a series of small and isolated populations collectively considered Threatened.

The main threat to the Canada Lynx is habitat loss, fragmentation and degradation by overly destructive forestry practices or deforestation. Pressure on habitat, poaching and roadkills are the main threats to the species in the US. Opening up of habitat also fosters the northwards movement of Coyotes and Bobcats, which may be an ancillary factor in suppressing Lynx numbers, for example in north-eastern US and eastern Canada. In most of the Lynx's Alaskan and Canadian range, habitat quality is high and habitat is intact, protected or relatively well managed. At least 11,000 Lynx are legally harvested per year, the great bulk of these in Canada and Alaska. Lynx populations are vulnerable to overharvesting during hare declines, though most legal hunting now takes this into account and there is little evidence of long-term population impacts. Climatic warming is already reducing the suitability of habitat for Lynxes along the southern periphery of the range and may have severe long-term impacts on boreal forest.

CITES Appendix II (ESA Threatened in the US). Red List: Least Concern. Population trend: Stable.

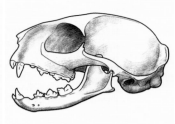

8.7–11.6cm

IUCN RED LIST (2015):
Least Concern

Head-body length ♀ 53–73.5cm,
Tail 27.5–59cm
Weight ♀ 3.5–7.0kg, ♂ 3.0–7.6kg

Jaguarundi

Herpailurus yaguarondi
(É. Geoffroy Saint-Hilaire, 1803)

Eyra

Taxonomy and phylogeny

The Jaguarundi's closest relative is the Puma with a common ancestor around 4.2 million years ago. Some authorities classify both in the same genus *Puma* though the Jaguarundi is generally accorded the unique genus *Herpailurus* given the considerable genetic distance and morphological differences between the two species. Eight subspecies are described, though recent molecular analysis of 44 wild Jaguarundis from nine countries shows little genetic differentiation across the range, and suggests most subspecies are probably invalid.

Description

The Jaguarundi has a very distinctive appearance with relatively short legs, a long, slender body and very long tail. The head is relatively small, elongate and flat with a characteristic blunt, 'Roman nose' profile and widely spaced, rounded ears. The Jaguarundi is the least marked of all small cat species, with short, sleek, uniformly coloured fur almost devoid of markings except for very faint facial stripes and highlights, and occasionally faint markings on the inner limbs; the ear-backs lack

spots. Kittens sometimes have spotting on the chest or belly but this is usually lost or indistinct by adulthood. The Jaguarundi has two distinct morphs, once considered separate species: iron-grey ranging from pale slate-grey to dark blackish-grey, and red-brown ranging from pale tawny to bright brick-red and often with a bright white muzzle and chin. The red-brown morph tends to be more common in dry, open habitats. Melanism is reported, though even very dark specimens are not entirely black and often have distinctly paler heads and throats. Litters can include kittens of both colours.
Similar species Distinct from all other neotropical cats. Its unusual appearance has been likened to a marten or otter (it is sometimes called Weasel-cat)

and it superficially resembles the Tayra, a neotropical mustelid. Tawny individuals are a similar colour to the much larger Puma.

Distribution and habitat
The Jaguarundi occurs from the eastern and western lowlands of northern Mexico, throughout Central and South America to south-eastern Brazil and central Argentina. Its presence in Uruguay is uncertain. The Jaguarundi formerly occurred in the US, restricted to extreme southern Texas with the last known record being a roadkilled animal near San Benito, Texas, in 1986. There is no evidence it naturally occurred historically in Arizona or Florida despite occasional reports. The Jaguarundi lives mainly in lowlands,

Dark coloured Jaguarundis often have paler fur on the head and neck as in this adult photographed in Panama (and compare with the individual on page 155). Coincidentally, the large mustelid the Tayra has very similar colouration, sometimes creating confusion in brief sightings.

typically up to 2,000m and inhabits the widest habitat range of any of the small neotropical cats. They occur in all kinds of dry and wet forest, savanna woodlands, wet subalpine scrub savannas, swamplands, semi-arid scrub, chaparral and dense grasslands. It has even been reported at 3,200m in cloud forest in Colombia. The Jaguarundi is tolerant of open habitats but avoids areas lacking cover. They occur in human-modified or recovering habitat provided there is dense cover and high rodent densities, for example in pasture-grasslands with shrubbery, old-field-scrub mosaics and secondary forest, as well as plantations of eucalyptus, pine and oil palm.

Feeding ecology

Jaguarundi diet is based mainly on analysis of scats and stomach contents with few direct observations of hunting. Most prey weighs less than 0.5kg with the most important category comprising small mammals such as cane mice, grass mice, rice rats and cotton rats. Larger prey up to and exceeding 1kg, including cavies, small opossums and lagomorphs, are fairly commonly killed, while armadillos are recorded occasionally. There is one record each of predation on a Common Marmoset, South American Coati and Pampas Fox; brocket deer remains have been found in scats, though this could have been carrion. After mammals, the most important prey types are birds, especially ground-dwelling or ground-foraging species like tinamous, quail and doves; and reptiles, particularly tegu and whiptail lizards. Iguanas and snakes, including venomous pit vipers, are also recorded rarely. Jaguarundis have been observed catching small fish from drying ponds. They readily take domestic poultry and will enter chicken coops.

Jaguarundis appear to be largely or exclusively diurno-crepuscular and forage mainly on the ground.

A Jaguarundi pauses in front of a camera-trap to investigate an interesting smell, Nairi Awari Indigenous Reserve, Costa Rica. Olfaction plays a relatively minor role in felid hunting and predatory behaviour, but it is central to their social lives.

Three collared Belizean cats were most active between 04:00 and 18:00 peaking at 11:00, while collared Jaguarundis in north-east Mexico were most active at 11:00–14:00. Camera-trapped Jaguarundis were only recorded between 05:30–06:00 and 18:00–18:30 during surveys in Atlantic Forest (Argentina and Brazil), dry savannas (Kaa-Iya del Gran Chaco National Park, Bolivia, and *caatinga* in Brazil) and *cerrado* grasslands (Brazil). This is consistent with most prey being diurnal and terrestrial. They are capable swimmers and can cross medium-sized rivers; for example, one was filmed swimming the Tuichi River, Madidi National Park, Bolivia, though there is no evidence they hunt in anything other than very small water bodies. Similarly, they are adept climbers and are presumably capable of taking prey in lower branches, though there is little evidence for arboreal hunting.

Social and spatial patterns

The Jaguarundi is poorly studied, with limited data from telemetered animals available only from Belize, Brazil and Mexico (only the last monitored a large number of animals, i.e. 21). Telemetry and camera-trapping show that Jaguarundis are primarily solitary despite a scattering of anecdotes of pairs being common (probably mating pairs or mothers with large kittens), but they are apparently gregarious in captivity. They appear to follow a fairly typical small felid socio-spatial pattern with archetypal felid marking behaviour, though range size of the sexes does not differ greatly and radio-tracked individuals overlap extensively. It is not clear if territories are defended from same-sex conspecifics. Mean home ranges in Mexico are slightly larger for females – 16.2km^2 (females) and 12.1km^2 (males) – but were generally smaller for females in a small number of monitored cats in Brazil

A female Jaguarundi photographed in oil-palm plantation, Colombia. Although Jaguarundis are found in agroforest habitat, their ability to inhabit large-scale plantations is poorly understood.

Right: **Essentially solitary cats such as the Jaguarundi actually inhabit a rich social environment in which individuals constantly exchange information with potential mates and rivals primarily by scent-marking.**

– 1.4–18km² (females) and 8.5–25.3km² (males). Range size for an adult female in Cockscomb, Belize was 20.1km² and unusually large ranges of 88–100km² were estimated for two young males that were likely dispersing. There are no reliable density estimates but they are captured at unexpectedly low frequencies compared to other cats during both camera-trap surveys and trapping efforts for radio-collaring, suggesting they are considerably less abundant than widely assumed.

Below: **A pale grey male Jaguarundi grooms a red female in captivity. Adult Jaguarundis do not form permanent groups in the wild but such bonding behaviour is bound to occur during mating associations (C).**

Reproduction and demography

Largely unknown from the wild. Anecdotal reports suggest they breed seasonally in some parts of the range though the evidence is weak. Captive females give birth to litters year-round. In captivity, oestrus lasts three to five days and gestation is 72–75 days. Litters number one to four kittens (average 1.8–2.3). Weaning begins around five to six weeks. Sexual maturity is 17–26 months. In the wild, kittens are kept hidden in dens located in dense vegetation and hollows in trees.

Mortality There are few records from the wild. Confirmed predators include Pumas as well as one record of a 2.7m Boa Constrictor killing an adult Jaguarundi in Central Mexico. They are vulnerable to domestic dogs near villages.

Lifespan Unknown in the wild, 10.5 years in captivity.

STATUS AND THREATS

Jaguarundis are considered widespread and they have a wide habitat tolerance including the ability to persist in some human-modified landscapes. However, it may be a misconception that they are generally considered common in most of their South American range. While their diurnalism and use of open habitats makes them more visible than other cats, this does not necessarily reflect abundance. There is some evidence they naturally occur at low densities especially in areas with abundant Ocelots, Bobcats or Coyotes, which may competitively exclude Jaguarundis. Threats are not well understood, but apart from the pervasive threat of habitat loss, the most important local threats appear to be fairly widespread persecution for killing poultry and frequent roadkill. They are seldom targeted for their unicolour skins but are killed by dogs and opportunistically by people (from which skins sometimes enter local trade). Jaguarundis in Central America are considered Vulnerable or Endangered, and the species is Critically Endangered if not already extinct in North America.

CITES Appendix I (Central and North America), otherwise Appendix II. Red List: Least Concern. Population trend: Decreasing.

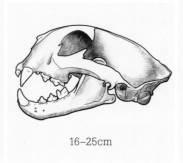

16–25cm

IUCN RED LIST (2008): Least Concern

Head-body length ♀ 95–141cm,
♂ 107–168cm
Tail 57–92cm
Weight ♀ 22.7–57.0kg, ♂ 39.0–80.0kg

Puma

Puma concolor (Linnaeus, 1771)

Cougar, Mountain Lion, Panther (Florida)

Taxonomy and phylogeny

The Puma is the only species classified in the genus
Puma. Its closest living relative is the Jaguarundi
with a common ancestor an estimated 4.2 million
years ago. More distantly, both are related to the
Cheetah and these three species together comprise
the *Puma* lineage.

   As many as 32 Puma subspecies have been
described, most of which are based on minor
morphological differences and are not valid.

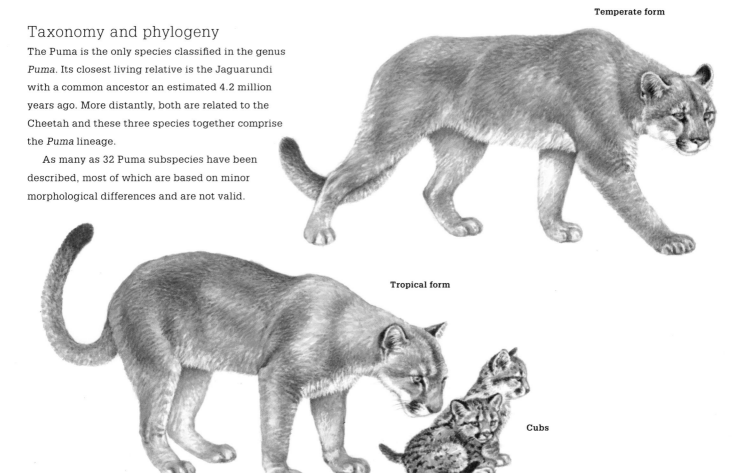

Temperate form

Tropical form

Cubs

Genetic analyses show three geographic groups, in North, Central and South America, with the most genetically diverse populations occurring in South America. On this basis, six subspecies are generally accepted: *P. c. couguar* (North America), *P. c. costaricensis* (Central America), *P. c. capricornensis* (eastern South America), *P. c. concolor* (northern South America), *P. c. cabrerae* (central South America) and *P. c. puma* (southern South America). The remnant population in Florida is usually classified as its own subspecies *P. c. coryi,* though its isolation occurred only in the last 100 years and genetic data indicate it is closely related to other North American populations. It is more correctly considered an ecologically and geographically

distinct population of the North American subspecies *P. c. couguar.*

Description

The Puma is the largest non-pantherine cat, comparable in size to the Leopard, with a robust head and neck, heavily built forequarters, slender hindquarters and muscular limbs. The long, tubular tail measures around two-thirds of head-body length and is distinctive in the field. The Puma's body size is broadly correlated with changes in latitude and therefore climate, available prey and possibly the presence of the Jaguar. The largest individuals occur in the temperate extremes of the range; for example, averaging 44kg (females) to 71kg (males) in Sheep

This young adult female Puma in Venezuela has the slender build of a subadult. Her rangy appearance is enhanced by having very short fur, typical of subtropical and tropical populations compared to their temperate counterparts.

Similar species The Puma was named for its resemblance to the female Lion and is widely referred to as the Lion in the western US and *León* in southern South America, but the resemblance is superficial and the two species are not sympatric in the wild. The Jaguarundi's uniform colouration and long tail might cause confusion in the Puma's Latin American range but the former is much smaller. Melanistic Jaguars should not be mistaken for Pumas given there are no verified records of black Pumas anywhere in the range.

This Puma in Torres del Paine National Park, Chile is fully protected but the species is heavily persecuted on sheep farms outside the park. Southern Patagonia has extremely high rates of anthropogenic killing of Pumas including by legal bounties on the Argentinean side.

Distribution and habitat

The Puma has the largest north–south distribution of any terrestrial mammal in the western hemisphere, ranging from the Yukon-British Columbia border in south-western Canada almost to the Straits of Magellan, Chile. It has a wide and mostly continuous distribution in south-west Canada, western US and tropical South America. It is relatively widely distributed in Mexico and Central America, though fragmentation and range loss are more advanced. In temperate Latin America, it occurs widely along a broad band following the Andes and extending into the Argentinean *pampas*, Argentine Monte, the Patagonian steppe and scrub, and throughout northern Argentina, Paraguay and northern Uruguay. It is extirpated from large areas of central-eastern and north-eastern Argentina, and central and northern Chile. It is extinct east of the Mississippi River in the US except for 100–120 in south Florida ('Florida Panther') and occasional dispersers into the north-east from the nearest breeding populations in the Dakotas.

River Wildlife Sanctuary, Canada; and 45.1kg (females) to 68.8kg (males) in Torres del Paine National Park, Chile. They are smaller in tropical regions; for example, averaging 36.9kg (females) to 53.1kg (males) in Iguaçu National Park, Brazil.

The Puma is uniformly coloured without body markings, typically light to dark tawny-brown with creamy-white underparts. Temperate Pumas tend to have paler, light greyish colouration particularly in the long, dense winter coat, while tropical individuals tend to rich, brick-red tones, but colour alone is not useful for distinguishing populations. The tail tip and the ear backs are dark brown to black, and the white muzzle is bordered by black. Despite persistent anecdotal reports, there are no records of complete melanism; a very dark mahogany brown individual with pale underparts and muzzle was killed in Guanacaste, Costa Rica. Albinism is recorded very rarely. Puma cubs are born with rich, dark brown spots and blotches covering the body, thought to be indicative of a more heavily spotted ancestral species, and perhaps useful for camouflage while Pumas are very young. The spots usually fade by 9–12 months, with rare cases of them being retained into adulthood.

The Puma occurs in a very broad range of temperate, subtropical and tropical habitats provided there is vegetation or rocky terrain. They inhabit all kinds of forest, woodland and scrubland; wet or dry, well-vegetated grassland savannas such as the Pantanal and *pampas*; and sparsely vegetated or rocky deserts. Pumas mostly shun open habitat, such as extensive grassland, prairie and barren desert, though they penetrate open landscapes in corridors and fragments with cover such as along watercourses, and they are able to cross large stretches of relatively open, marginal habitat. They are tolerant of human proximity, though they do not permanently occupy heavily modified landscapes such as croplands and monoculture plantations. They live close to people, including in rural and peri-urban areas where there is suitable habitat with prey. The Puma occurs from sea level to around 4,000m in North America and up to 5,800m in the Andes.

Feeding ecology

A Puma at its Guanaco kill, Torres del Paine National Park, Chile. Pumas in this ecosystem kill an average of 6.5 animals a month, most of which are large ungulates.

The Puma is recorded eating a wide range of species from arthropods to adult male Elk (~400kg), but the diet's mainstay comprises medium-sized to large mammals. Large kills are more common in temperate regions particularly in North America where Pumas have access to abundant populations of large ungulates; the Bison is the only potential species invulnerable to Puma predation. White-tailed Deer, Mule Deer and Elk are the main prey species in Canada and the US. Bighorn Sheep, Moose (mainly juveniles), Pronghorns, Mountain Goats (rarely) and Collared Peccaries are also killed and are often locally important when abundant. Collared Peccaries, White-Tailed Deer and feral pigs are the major prey in the south-western US, Florida and northern Mexico. Bighorn and Mule Deer are the two main prey species for Pumas in the northern Rocky Mountains, Alberta, and Bighorn are the most important prey species for Pumas in north-western Sonora, Mexico. In the southern temperate range (Patagonia), the diversity of the native prey community is particularly low and Pumas focus heavily on the only large ungulate present, the Guanaco. In central Chilean Patagonia, Guanacos

comprised 88.5 per cent of 463 recorded kills of seven species (including domestic sheep), and were 59 per cent of biomass eaten by Pumas in Torres del Paine National Park, southern Chile, during a period when Guanaco populations were recovering. In subtropical and tropical parts of the Puma's range, ungulates remain an important component of the diet but less so given that ungulate abundance and biomass declines, and the diet is more diverse and prey size is smaller. On average, ungulates comprise around 35 per cent of prey items in Latin America compared to 68 per cent for North America. Important prey species for tropical Pumas include Red Brocket Deer, Grey Brocket Deer, Collared Peccaries, White-lipped Peccaries, Capybaras, Spotted Pacas, agoutis and armadillos. Across the range, Pumas incidentally feed on a wide variety of small mammals such as lagomorphs, porcupines, marmots, beavers, small rodents, anteaters, sloths, marsupials and primates. Locally or seasonally, such prey may be an important supplement to the diet. Introduced European Hares are abundant in Torres del Paine, Chile, and Pumas eat an estimated 13 hares for every Guanaco in parts of the park where Guanacos occur at high densities. In ranching areas of north-western Argentine Patagonia (Neuquén province, Argentina), native prey is now rare and introduced mammals comprise 99 per cent of the diet by mass, mainly European Hares (45 per cent of biomass) and European Red Deer (43 per cent). The European Red Deer is expanding rapidly along the southern Andes throughout Patagonia and is becoming increasingly important in the Puma's diet there.

Pumas are recorded killing at least 30 species of other wild carnivores including Canada Lynx, Bobcats, Ocelots, Jaguarundis, Geoffroy's Cats,

The cooler temperatures of temperate ecosystems slow down carcass decomposition, allowing Pumas to remain at large kills for many days. This Puma has almost entirely consumed a Bighorn Sheep, Glacier National Park, Montana.

A young Puma feeds on its mother's kill of a Mule Deer. The kill has been partially buried with leaf-litter to reduce the chances it attracts scavengers, particularly those dangerous to cubs such as bears and wolves.

most often in the diet of tropical Pumas but at low levels, rarely exceeding 5 per cent of diet. Pumas kill livestock, mostly free-ranging sheep, goats and young cattle. Livestock rarely forms a major part of the diet and Pumas cause relatively little damage over most of their range, especially in North America where losses tend to be localised and temporary. However, livestock may dominate the diet where it is the most abundant option (for example, cattle in the Brazilian Pantanal) or where wild prey has been reduced (for example, sheep in much of Patagonia). Feral livestock forms an important part of the diet in some Puma populations, for example wild boar in Florida and wild horses in Nevada. Pumas occasionally take domestic dogs and cats in peri-urban areas. Humans are very rarely killed, though fatalities are usually considered predation rather than defence. There are only 23 recorded fatalities in North America (Canada and US) from 1890 to 2012, two of which were deaths from rabies after an attack by a rabid Puma.

Pumas hunt chiefly at dusk, night and early morning. Daylight hunts occur more often in winter, in areas where Pumas are not persecuted, and are more likely to be opportunistic rather than the result of a concerted search for prey. Hunting Pumas search for prey by travelling along key features of the landscape such as animal trails, logging roads, watercourses, valley bottoms and ridgelines. Active searching is interspersed with resting and lying in wait near areas where prey congregate such as at water sources and feeding sites like meadows. A Puma in Patagonia repeatedly swam 549–1,087m to a lake island where he killed sheep and swam back to the mainland each time within 24 hours. Pumas stalk very close to prey, typically within 10m of large ungulate quarry, before an explosive final rush. Most successful hunts have very short chases, within 10m of initiating the hunt in the case of Pumas attacking Mule Deer in south-eastern Idaho and north-western Utah. Large prey is usually subdued quickly at the site of contact or within 10–15m if a struggle ensues. There are numerous accounts of prolonged struggles

Colocolos, Coyotes, Grey Wolves, Maned Wolves, foxes, raccoons, coatis, mustelids and skunks. Carnivores may or may not be eaten and are rarely important prey, though Northern Raccoons are around 20 per cent of Puma diet by biomass in Florida. Cannibalism is recorded. Pumas opportunistically kill a wide variety of birds ranging from sparrows to rheas, and reptiles up to the size of adult caimans and American Alligators, though these contribute little to intake. Birds and reptiles feature

with very large prey over distances up to 80–90m, and, in some cases, Pumas have been carried along on the back of species such as Elk before bringing it down. Ungulates are usually killed by asphyxiation with a throat bite while smaller prey is bitten in the skull or nape.

Hunting success of Pumas is surprisingly poorly known for such a well-studied species. Taking only actual attempts when an attack was launched, 82 per cent of hunts of deer and Elk by Pumas in Idaho were successful; this figure excludes stalking attempts that were aborted before an attack. Pumas drag kills up to 80m into vegetation, and may cover them with dirt, leaf-litter or snow to help preserve them and hide them from scavengers. They return to large kills repeatedly, consuming the carcass over a period of three days to four weeks for very large kills in winter. Pumas scavenge, though this usually represents a small amount of their intake. In contrast, Pumas often leave carcasses before they are entirely consumed; each individual Puma in central Chilean Patagonia abandons an average of 172kg of edible meat per month to scavengers. Grey Wolves, Grizzly Bears and Black Bears drive Pumas from kills in western North America.

Social and spatial behaviour

Pumas are solitary and territorial. Both sexes maintain enduring territories in which the ranges of males are large and typically overlap one or more, smaller ranges of females. Adults defend their territories against same-sex conspecifics and fights over territory are occasionally fatal, particularly for males. Nevertheless, non-violent social interactions

Most sightings of Puma groups are explained as a mother with large cubs, such as this pair in Torres del Paine National Park, Chile. The nearly adult-sized cub (standing) is close to the age at which she will disperse.

between adults are more common than is generally believed. GPS-collared adults in the Greater Yellowstone Ecosystem interact fairly regularly with little or no aggression. As well as encounters between males and females during the mating season, adults (especially females) share large kills, and interactions between unrelated adults are as frequent as between related adults. Intrasexual territorial overlap varies between populations; in general, males overlap little and females overlap more, in some cases extensively. Overlap is least for both sexes and particularly males where range size is small.

Territory size varies with habitat quality and prey availability, from 25 km² (female, Venezuela) to 1,500km² (male, Utah). Territory size is best quantified from North America where representative

estimates of the mean range size include 74km² (females) and 187km² (males) in an unhunted population in New Mexico; 140km² (females) and 334km² (males) in a hunted population in Alberta; and 191km² (females) and 558km² (males) in the unhunted population in Florida. Puma socio-spatial ecology is less well studied in Latin America but limited information from subtropical and tropical habitats suggests that range size is smaller and densities are higher. In dry forest habitat in Venezuela range size averages 33km² for females, and 60km² for males based on a very small sample and estimated only for the dry season; range size is believed to increase during the wet season. In tropical dry forest, northern Mexico, range size increased from 25km² for females in the dry season to 60km² in the wet season; and 60km² to 90km² for

males (again based on very few data). Ranges in Pantanal-*cerrado* habitat, Brazil, average 89km² (females) and 222km² (males).

Density of northern Pumas typically varies from 0.3 Pumas per 100km² (for example, Utah, Texas) to 1–3 Pumas per 100km² (Alberta, California, Utah, Wyoming). Puma density exceeds three per 100km² in exceptional circumstances, for example up to seven Pumas per 100km² on Vancouver Island, British Columbia, which has very high densities of non-migratory deer prey. Density estimates from Latin America include 0.5–0.8 Pumas per 100km² (subtropical forest, Argentina); 2.5–3.5 Pumas per 100km² (grassland steppe, Patagonia); 2.4–4.9 Pumas per 100km² (tropical rainforest, Belize); and 3–4.4 per 100km² (Pantanal seasonally flooded savannas, Brazil).

Reproduction and demography

The Puma breeds year-round and litters can be born in any month, though seasonal birth pulses are prevalent in some populations, typically coinciding with birth periods of ungulate prey and to avoid seasonal extremes (especially of winter). Around 75 per cent of litters in the northern Rockies (Canada and US) are born from May through October, and 50 per cent of litters in Yellowstone National Park, US, are born in June and July. In Florida, the birth peak occurs slightly earlier, from March to June, linked to White-tailed Deer births. Patterns are poorly known for Latin America although reproduction is likely to be the least seasonal in tropical habitats. Oestrus lasts 1–16 days and gestation lasts an average of 92 days (range 82–98 days). Litter size is usually one to four cubs and averages two to three; up to six cubs are recorded in captive litters. Weaning occurs around two to three months. Adoption occurs rarely; for example, an adult Wyoming female with three six-month-old cubs adopted two 15-month-old male orphans; it is likely the female was related to the orphans' mother. Inter-litter interval is 14–39 months, averaging 26.5 months (Wyoming). Both sexes are sexually mature at around 18 months. Wild

females usually first give birth after 24 months and exceptionally at 18 months old; males first breed at around three years.

Cubs reach independence from their mothers on average at 13.5–18 months (range 10–24 months) and dispersal typically occurs a few months after independence. Both sexes disperse. Females are philopatric when possible and are more likely to settle close to their natal range; female immigration is common in areas where Pumas are heavily hunted. Male philopatry is exceptional and has only been recorded in the isolated Florida population. Females usually disperse shorter distances and settle more quickly than males. Florida females disperse 6–32km (mean 20.3km), males 24–208km (mean 68.4km); San Andreas Mountains (New Mexico) females disperse 0.7–79km (mean 13km), males 47–215km (mean 68.4km). Dispersal in two mountainous sites in Utah followed the usual pattern in Pumas of female dispersal being less common than in males, although dispersal distance was similar for the sexes, and the largest distances moved were by females, 11–357km (mean 33–65km) compared to males 6–103km (mean 31–52km). The longest dispersal distances are recorded in the isolated Black Hills, South Dakota, population, which is surrounded by prairie and agriculture; females disperse 12–99km (mean 48km) and males 13–1,067km (mean 275km). In 2011, a dispersing male from this population moved a straight-line distance of approximately 2,800km before being hit by a car in coastal Connecticut.

Mortality Estimates of annual adult mortality include 9 per cent (males) to 18 per cent (females) in New Mexico; and 35 per cent (females) to 61 per cent (males) in north-west Montana. Cub survival to independence is estimated at 45–52 per cent (southern California) to 64 per cent (New Mexico). An estimated 50 per cent of cubs survive to independence in Yellowstone National Park, but this figure drops to 21 per cent when calculated from birth to two years old. Adult Pumas are mostly killed by people, mainly by legal hunting (in North America) and illegal killing. Mortality from vehicles

Opposite: **A Puma mother and her three cubs rest in front of a cave in National Elk Refuge, Wyoming. Female Pumas have recently been documented bringing small, live prey to their cubs, providing them with opportunities to learn how to handle prey.**

Unless the den is disturbed, mother Pumas rarely move cubs from the birth site for the first two months of life (C).

is a significant factor for some populations, notably in Florida where it is the main cause of death (54 of 88 known deaths 2010 to June 2013). Natural factors from which adult Pumas die include starvation, disease and hunting accidents. Very high rates of mortality from other Pumas were recorded in one New Mexico population where 46 per cent (for males) to 53 per cent (for females) of deaths were attributed to adult males killing other adults. However, this was either an extremely unusual situation or the result of circumstantial evidence. There is no evidence of such high rates of intraspecific killing from other well-studied Puma populations. Cubs die mainly from starvation (including when mothers are killed in legally hunted populations), predation (especially by Grey Wolf where the two species are sympatric) and infanticide.

Lifespan Up to 16 years in the wild and up to 20 years in captivity.

STATUS AND THREATS

The Puma is widespread, resilient and is relatively tolerant of human presence and activity. Large areas of its distribution in North and South America are still essentially contiguous where it is generally considered secure. Although the species still has the largest distribution of any terrestrial mammal in the western hemisphere, it is extinct in its entire eastern North American range except for the relict Florida population, and it has been extirpated from at least 40 per cent of its Latin American range. In North America, it may be gradually reclaiming some of its former range in the Midwestern US and Canada with 178 confirmed records (mainly dispersing males) outside known breeding range in 1990–2008. In contrast, fragmentation of habitat by growing human populations threatens some current strongholds, such as in southern California, and limits colonisation by isolated populations, for example in the Dakotas and Florida. The Puma's status is poorly known in much of Latin America. It is likely to be widespread and common where large areas of habitat remain but more tenuous elsewhere, for example in much of Central America where pressure on habitat is more extreme. It was extirpated from much of its temperate Latin American range but has recovered or is reclaiming much of this in the last 20–30 years, including most of the Patagonian steppe and scrub, and large parts of the humid Argentine *pampas*.

Pumas are threatened chiefly by habitat loss and fragmentation, and the associated threats brought by people. There is often intense persecution in livestock areas, especially in Latin America where killing of Pumas by people is widespread and indiscriminate. Southern Argentina is the only region that still pays bounties (since 1995) to kill Pumas in sheep-ranching areas, resulting in extremely high rates of removal, approximately 2,000 Pumas per year; a further 80 Pumas are legally sport-hunted there. Legal sport-hunting kills 2,500–3,500 Pumas per year in the US and Canada. Sport-hunting is known to trigger population declines when excessive quotas are granted, especially of adult females. Overhunting of Puma prey species by people is a possible threat in much of its tropical range where subsistence hunters target the main prey species of large cats.

CITES Appendix I (Florida, Nicaragua through Panama), elsewhere Appendix II, legally sport-hunted in Argentina, Canada, Mexico, Peru and US (hunting is not permitted in California or Florida). Red List: Least Concern. Population trend: Decreasing.

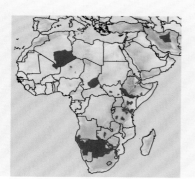

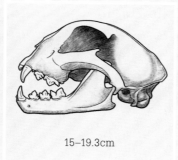

15–19.3cm

IUCN RED LIST (2008):
- Vulnerable (global)
- Critically Endangered (Asia)
- Critically Endangered (*hecki* subspecies)

Head-body length ♀ 105–140cm, ♂ 108–152cm
Tail 60–89cm
Shoulder height ♀ 67–89cm
Weight ♀ 21–51kg, ♂ 29–64kg

Cheetah

Acinonyx jubatus (Schreber, 1775)

Taxonomy and phylogeny

The Cheetah is the sole species in the genus *Acinonyx*. With its closest living relatives, the Puma and Jaguarundi, it is part of the *Puma* lineage, which diverged around 6.7 million years ago. The oldest Cheetah fossils are 3–3.5 million years old from southern and East Africa.

Four African and one Asian subspecies are currently described. The Asiatic cheetah *A. j. venaticus* is the most distinct subspecies, with genetic differences indicating isolation from African populations for an estimated 32,000–67,000 years. Cheetahs in north-eastern Africa (traditionally *A. j. soemmeringii*, the Central African Cheetah, Somalia to eastern Niger) are genetically distinct from other African populations, though they are likely contiguous with the West African cheetah (traditionally *A. j. hecki*, western Niger to Senegal).

Saharan form

Typical form

Cub

Right: **The Cheetah is less sexually dimorphic than most large cat species. Nonetheless, adult males are significantly more heavily built than females, especially in the head, neck and forequarters.**

Further molecular research may combine them as one subspecies. Similarly, African Cheetahs north of the Sahara were formerly considered *A. j. venaticus* but a western Egypt sample shows they are not closely related to Iranian cheetahs, and are likely to cluster with *soemmeringii/hecki*. Cheetahs from southern Africa (*A. j. jubatus*) and East Africa (*A. j. raineyii*) are closely related, with relatively minor genetic differentiation, though they are presently considered two subspecies.

Description

The Cheetah is the only felid adapted for prolonged high-speed pursuits of prey, with an unmistakable, greyhound-like build unlike any other cat. It has a tall, slim frame with long legs, a narrow, deep chest, narrow waist and long, tubular tail. The head is small and rounded, with a short muzzle and small ears. Cheetah claws are dog-like and they lack the

Below: **One of the last remaining Asiatic Cheetahs, photographed by camera-trap in Naybandan Wildlife Refuge in Iran. This adult male shows the very short coat typical of Asiatic Cheetahs during the extremely hot Iranian summer.**

fleshy, protective claw-sheaths present in other cats, but, contrary to popular belief, the claws are partially protractile. Claws appear in the tracks except for the sharp, strongly curved dew claw that is used for prey capture. Southern African cheetahs tend to be largest, closely followed by those from East Africa. The smallest Cheetahs (based on very few measured specimens) are from the Sahara, north-eastern Africa and Iran.

The Cheetah is typically light tawny to yellow-blonde fading to creamy-white underparts. About 2,000 solid black round or oval spots cover the body, interspersed sporadically with small, often indistinct black dabs. The pattern these markings make is unique to each individual. The Cheetah's lachrymal or 'tear' streaks are not found in any other felid and their function, if any, is unclear – they may help to reduce glare for diurnal hunting and enhance defensive or aggressive facial expressions. Saharan Cheetahs tend to have distinctive narrow, dog-like faces and short, pale fur ranging from yellow-beige with chocolate-brown spots to near-white with faint cinnamon or ochre markings. The so-called King Cheetah was considered a separate subspecies *A. j. rex* when first described in 1927 but is merely a recessive colour variant which may be born to normally spotted parents. In the wild, the king variant appears regularly only in northern South Africa (including Kruger National Park), southern Zimbabwe and south-eastern Botswana. A single king skin was confiscated from a poacher in Burkina Faso, possibly a trade skin. Melanism is extremely rare, with only two confirmed specimens, from Zimbabwe and Zambia. There are no confirmed records of true albino Cheetahs, but a yellow-cream-coloured Cheetah with dark freckling was photographed in Kenya's Athi River region in 2010.

Uniquely among cats, Cheetah cubs are born with a fluffy, smoky-grey mantle that runs from the crown to the base of the tail. The mantle dwindles by four to six months to become a short mane on the shoulders that is inconspicuous in most adults

except when raised in fear or aggression. The mantle is more prominent in Asiatic Cheetah adults, especially during winter when Iranian individuals develop long, silky coats. The function of the mantle is unclear. A popular theory suggests it mimics the irascible Honey Badger to deter predators. Although the similarity is striking, its value is dubious given the very high rates of predation on cubs in some populations. More likely, the mantle probably assists with thermoregulation and perhaps in camouflaging young, helpless cubs in long-grass dens.

Similar species The Cheetah is similar in overall size and colouration to the Leopard, but confusion is unlikely, except perhaps in fleeting sightings – the Cheetah's simple spots and tear streaks are distinctive. Servals are often identified by local people as Cheetahs, though they are much smaller with a characteristic short tail.

Distribution and habitat

The Cheetah is relatively widely distributed in southern and East Africa, rare to very rare in West and Central Africa and extinct in North Africa except

Once considered a separate species, the King Cheetah is simply the equivalent of a tabby-coloured Cheetah. A mutation in the same gene that produces tabby markings in domestic cats causes the spots to coalesce into blotches and stripes.

southern Algeria and perhaps western Egypt. It is extinct in Asia except for approximately 50 animals in central Iran. Occasional reports since the 1970s from Afghanistan, Pakistan and Turkmenistan lack unequivocal evidence. A skin found in Mazar-e-Sharif in 2006 reportedly came from Samangan Province, northern-central Afghanistan – it is more likely, however, to have originated in Iran.

The Cheetah favours savanna woodland, grasslands and scrublands, reaching its highest densities in mesic mosaics of open woodland and grassland. It is more sparsely distributed in dense, humid woodlands such as Miombo, and does not occur in thick forest and rainforest. They are well adapted to arid savannas such as in the southern

Kalahari, and they inhabit semi-desert and desert in Iran, the Namib and the Sahara where they are associated with watercourses and mountain ranges; they are transient in the driest areas of true desert. Cheetahs are recorded exceptionally to 3,500m (Mount Kenya, Kenya) but suitable habitat for Cheetahs typically dwindles at around 1,500m (Ethiopia) to 2,000m (Algeria). Iranian Cheetahs live above the winter snowline in desert massifs, the only place where the species routinely experiences winter snow.

Feeding ecology

The Cheetah is famously the fastest terrestrial animal on Earth, best suited for hunting small to medium-sized antelopes weighing 20–60kg; gazelles or gazelle-analogues are preferred throughout the range. Typical primary prey species include Thomson's Gazelle and Grant's Gazelle (East Africa), Dama Gazelle and Dorcas Gazelle (Sahara), Springbok (arid southern savannas, e.g. Etosha National Park and the Kalahari), and Impala (southern and East African woodlands, e.g. Kruger National Park, Okavango, Serengeti woodlands). Smaller antelopes such as Steenbok and Common Duiker are important prey, for example in the southern Kalahari. Cheetahs are capable of taking large ungulates, which may constitute preferred prey in areas where they are the most abundant option, e.g. Nyala (adults 55–127kg) in mesic woodland, Phinda Game Reserve, South Africa, or where smaller prey is unavailable, e.g. Urial sheep (adults 36–66kg) and Wild Goat (adults 25–90kg) in Iran following the widespread extirpation of gazelles. Generally, juveniles are preferred prey over adults and females are preferred over males, especially when prey is large or dangerous, such as Warthogs. Male gazelles are often seasonally preferred where they have a defined breeding season as males compete with other males making them less vigilant and more vulnerable to Cheetah predation, e.g. Springboks in the Kalahari. Cheetahs, typically male coalitions, readily kill juveniles of large species including Blue

Provided they survive, these young brothers will remain together as a male coalition once they become independent. The strong bonds they form as cubs will endure for life, with almost constant contact, cooperative behaviour and affectionate displays.

Wildebeest, Burchell's Zebra, Gemsbok and, rarely, African Buffalo and Giraffe. Male coalitions are also capable of taking large, dangerous prey that individual male and female Cheetahs usually avoid, for example adult Hartebeest, oryx and wildebeest.

Large rodents, such as Spring Hares, and large lagomorphs (Cape Hares, Scrub Hare) are important prey in some situations, for example for single Kalahari Cheetahs, for recently independent young adults in many populations, and in impoverished habitats, e.g. in Iran. Cheetahs sometimes kill ostriches, bustards and guineafowl, and rarely hedgehogs (Sahara), porcupines and small carnivores including mongooses and foxes. Cannibalism has been recorded by males following

fatal, territorial clashes but is otherwise unknown. Cheetahs in the Kalahari reputedly eat tsamma melons for the water content but a comprehensive six-year observational study did not record this, and it is unlikely tsammas contribute much if anything to water intake. Cheetahs kill small, untended livestock, such as goats, sheep, calves and juvenile camels, but they are easily deterred by people or dogs. There is no record of a wild Cheetah killing a human.

In the majority of their range, Cheetahs are most active during the day, maximising visibility and reducing overlap with peak activity periods of Lions and Spotted Hyaenas. Most hunts occur within a few hours of sunrise and sunset, but often throughout

Killing by suffocation is a safe technique for subduing potentially dangerous prey. This Cheetah killing a young Blesbok shows how cats further reduce the chance of injury from struggling prey by immobilising the horned head and positioning themselves away from thrashing hooves.

the day during cooler periods and if females have cubs. Nocturnal activity is more common than widely believed. About 25 per cent of all activity by Okavango cheetahs occurs during the night, mostly around full moons; nocturnal hunting occurs but it is unclear if 25 per cent of activity translates to 25 per cent of hunting effort. Saharan Cheetahs are thought to hunt frequently at night, probably taking advantage of cooler temperatures and very open habitat. Cheetahs usually initiate a hunt by stalking in fairly typical felid fashion, semi-crouched and freezing when prey looks up, to within 50–75m of the target. They may, however, start trotting openly towards very vulnerable or oblivious prey, such as neonate antelopes, from 500 to 600m away. Sprints last up to 600m but typically less than 300m; chases average 173m for Cheetahs hunting in open savanna woodlands (Okavango, Botswana). A trained cheetah running behind a vehicle was timed at 105kmph, the fastest recorded speed – the top speed recorded for wild Botswanan Cheetahs is 93kmph. The Cheetah is probably capable of exceeding 105kmph, but adults hunting mainly Impalas in Botswanan savanna woodlands use moderate speeds, with a top speed averaging 54kmph. Cheetahs hunting gazelles in very open habitat probably average faster hunting speeds. The Cheetah can accelerate and decelerate faster than any terrestrial mammal. Cheetahs decelerate very rapidly in the final stage of the hunt which enhances manoeuvrability for making the kill, e.g. from 58kmph to 14kmph in three strides (Okavango, Botswana). Prey is tripped or bowled over with the front paw, or pulled off balance by hooking with the dew claw and yanking backwards while decelerating. Most prey is killed by suffocation, taking 2–10 minutes, but hares and other small prey are killed rapidly by a bite to the skull or nape.

Around 25–40 per cent of hunts are successful, with higher success rates for hunts of small and vulnerable prey: 87–93 per cent for hares, 86–100 per cent for juvenile gazelles (Serengeti National Park, Tanzania). Cheetahs rarely defend their kills, losing up to 13 per cent (Serengeti National Park), chiefly to

Cheetahs are poorly built for arboreality but they easily scale large sloping trunks, sometimes as high as 5–6m. Cheetahs value trees as communal signposts for scent-marking, and for scanning the horizon for prey.

Spotted Hyaenas and Lions, less often to Leopards, Wild Dogs, Grey Wolves (Iran), Brown Hyaenas and Striped Hyaenas. Groups of Black-backed Jackals and large flocks of vultures are recorded usurping Cheetahs from their kills but they probably leave fearing that larger carnivores will be attracted by the commotion. Cheetahs very rarely scavenge, presumably because of the danger of encountering larger carnivores, but male Cheetahs readily take kills from female Cheetahs. Cheetahs also seldom revisit previously fed-upon carcasses for the same reason, but mothers with cubs sometimes return to a kill after abandoning it overnight (Phinda).

Social and spatial behaviour

The Cheetah has a fluid and complex socio-spatial system unique among felids, with greater sociality than all other cats except the Lion (and some feral

domestic cat colonies). Female Cheetahs are asocial and non-territorial with large home ranges that are not defended. Encounters between females typically result in both actively avoiding contact or, rarely, resting together tolerantly for less than a day (often because they are related; see below). In contrast, males are social and often form lifelong coalitions of two to four individuals that compete aggressively with other males for access to females. Coalitions usually comprise brothers who remain together after dispersal. Pairs or singletons, however, often recruit an unrelated member; for example, 30 per cent of Serengeti coalitions contain a non-relative. Some males, particularly those without male siblings, may never belong to a coalition – around 40 per cent of Kalahari males are solitary. Males may or may not be territorial; because females are semi-nomadic, males typically establish territories only where it increases

access to females, for example when female ranges are small or their movements are predictable. Serengeti males establish small territories often centred on areas of cover and where Thomson's Gazelles may aggregate seasonally. Female Cheetahs follow gazelle migrations and are attracted to the aggregations. In areas with non-migratory prey (e.g. southern Kruger National Park, South Africa), female ranges tend to be smaller with less nomadic movements, making male territoriality profitable. Male territories in this scenario are a similar size to female ranges. Both coalitions and single males may be territorial, but coalitions are more successful at defending territories and therefore more likely to gain access to females. Serengeti coalitions are also healthier than singletons. Territorial males repel rival males in fights, which are sometimes fatal.

This wildebeest calf is no match for a Cheetah but its family is. Wildebeest and zebra herds often repel attacks on their juveniles by lone Cheetahs. They are much less likely to succeed in defending youngsters against Cheetah male pairs or trios.

Right: **Female Cheetahs with cubs spend much of their day scanning the environment. Vigilance by mothers is geared mainly towards finding prey except while feeding on kills, when females are keenly alert to the approach of other predators.**

As an alternative to territorial residency, males may be non-territorial, semi-nomadic 'floaters', either because they are unable to defend a territory or possibly because it is an equally successful strategy for finding females. All male Cheetahs start out as floaters and some never become resident; those that do may switch between territoriality and floating, depending on the availability of prey (and therefore of females) and whether they belong to a coalition. Floaters make up 40 per cent of the Serengeti population. All central Namibian males appear to be floaters with very large ranges, perhaps because of low prey densities and high human mortality creating constant disturbance in the population.

Range size is among the largest recorded for felids. Where prey is migratory or scarce, female range size averages 833km^2 (range 395–1,270km^2; Serengeti National Park) to 2,160km^2 (range 554–7,063km^2; central Namibia). Female ranges are smaller in habitat with resident or abundant prey, e.g. 34–157km^2 (woodland, Phinda Game Reserve) and 185–246km^2 (woodland, Kruger National Park).

Below: **If caught, this Cheetah would be no match for the Lioness but adult Cheetahs easily outrun other carnivores provided they are not surprised. The risk of predation for them increases in dense woodland habitat where visibility is reduced.**

Male territory size averages 33–42km² (single males and coalitions, Serengeti) and 93km² (range 57–161km²; coalitions, Phinda Game Reserve). In Kruger National Park, 126km² (three-male coalition) and 195km² (single male) were recorded. Floater males have very large home ranges, averaging 777km² (Serengeti National Park) to 1,390km² (range 120–3938km²; single males, central Namibia) and 1,464km² (range 555–4,348km²; coalitions, central Namibia). An Iranian male pair (likely floaters) used 1,737km² in five months of radio-tracking. Cheetahs naturally occur at low densities, generally 0.3–3 per 100km²; estimates include 0.16 per 100km² (Iran), 0.25–2 per 100km² (Namibian farmlands), 0.5–2.30 per 100km² (Kruger National Park, South Africa), 2 per 100km² (Serengeti National Park, Tanzania) and 4.4 per 100km² (Kgalagadi Transfrontier Park, South Africa). Temporary densities during migratory prey aggregations exceptionally reach 20 per 100km² (Serengeti National Park, Tanzania).

Reproduction and demography

Cheetahs have a reputation for poor reproduction arising from the difficulties of breeding them in captivity, but wild cheetahs are prolific. Although Cheetahs have relatively low genetic variation, there is no evidence of reproductive inbreeding depression in wild Cheetahs. Cheetahs breed year-round, sometimes with weak birth peaks during prey lambing periods, for example November to May, Serengeti National Park. Iranian Cheetahs purportedly mate in winter (January to February) and give birth in spring (April to May), though there are few confirmed records. Oestrus is one to three days and mating is typically feline, though extremely rare to witness in the wild, often taking place in thick cover and at night. Gestation is 90–98 days. Litter size is typically three to six, rarely to eight; a litter of nine from Kenya may have included adoption. Inter-litter interval averages 20.1 months (Serengeti). Weaning begins at six to eight weeks and is complete by four to five months. Cubs reach independence at 12–20 months (average 17–18

months) and disperse as a sib-group. Females leave the group before sexual maturity while the males remain together. Dispersing females usually settle near their natal range so that neighbouring females are often related. Males typically disperse further to avoid breeding with female relatives. Females can conceive at 21–24 months, first give birth around 24 months, averaging 29 months (Serengeti), and can reproduce up to 12 years. Males are sexually mature at 12 months, though rarely breed before three years.

Mortality 95 per cent of Serengeti cubs die before independence, most killed by Lions and less so Spotted Hyaenas, and most before eight weeks while cubs are still in the den. These are exceptionally high figures, and are likely restricted to the Serengeti's very open plains and were recorded at a time when Lion densities were high. Cub mortality is lower everywhere else – for example, 37 per cent of Kgalagadi Transfrontier Park cubs survive to independence, and most mortality is due to predation (as for the Serengeti). Post-emergence (i.e. after the vulnerable denning period) survival estimates include 67 per cent (Kgalagadi

Cheetah cubs play a game of 'king-of-the-hill'. Baby Cheetahs belie their delicate appearance and, like all young cats, engage in very energetic and vigorous play.

Transfrontier Park), 62 per cent (Phinda Game Reserve), 57 per cent (Nairobi National Park, Kenya), 50 per cent (Kruger National Park) and 20 per cent (Serengeti). Infanticide by male Cheetahs has never been recorded, perhaps because the wide-ranging movements of females limit the benefits to males of killing unrelated cubs. Adult Cheetahs are occasionally killed by Lions, Leopards and Spotted Hyaenas, and territorial fights are a significant cause of mortality for males. Hunting accidents are rare but are sometimes fatal: a female pursuing Impalas in dense woodland was disembowelled while clearing a tree stump (Londolozi Game Reserve, South Africa) and a Cheetah hunting a male Grant's Gazelle in the Serengeti died from injuries inflicted by the animal's horns. Despite high genetic homogeneity, disease is rare in wild Cheetahs.

Lifespan Maximum 14 (average 6.2) years for Serengeti females; 11 (average 5.3) years for males; 21 in captivity.

STATUS AND THREATS

The Cheetah has disappeared from approximately 80 per cent of its historical range in Africa and its entire Asiatic range except for a single population of around 50 in central Iran. Total numbers are estimated at 7,000–9,300 adults and independent adolescents. Southern Africa (with roughly 4,500–5,000) and East Africa (with roughly 2,600) are the main strongholds; Namibia (~2,000), Botswana (~1,800) and Tanzania–Kenya (~1,700 combined) have the largest populations. Cheetahs are extinct or Critically Endangered in North and West Africa; fewer than 250 adults are estimated in two populations in Algeria and the protected WAP Parks Complex (Niger, Benin and Burkina Faso). They are otherwise thought to be extinct or relict across the Sahara and Sahel except for relict populations in southern Chad and southern Sudan.

Conversion of habitat to farmlands with replacement of prey by livestock is a key factor driving Cheetah declines. Across their range, a majority of Cheetahs occur outside protected areas where habitat conversion is prevalent and accelerating. Cheetahs are widely persecuted by livestock farmers despite causing relatively minor damage, and they are profoundly affected by loss of prey in pastoral areas. Human hunting of prey is a severe threat in the Sahel, North Africa and Iran where cheetahs are naturally very rare. Limited hunting for skins occurs, for example across the Sahel into southern Sudan where the market is mainly for luxury items such as traditional footwear called *markoob*. There is also significant illegal trade in live cubs and adults in north-eastern Africa, primarily from Somali ports to the Arab Gulf States where there is a high demand for Cheetahs as pets. Most cubs in this trade die in transit. High genetic homogeneity has had little impact on wild populations. The Cheetah is legally sport-hunted in Namibia (2013 quota: 150) and Zimbabwe (2013 quota: 50).

CITES Appendix I, permitting trade of 205 hunting trophies and live animals. Red List: Vulnerable (global), Critically Endangered (Asia), Critically Endangered (*hecki* subspecies). Population trend: Decreasing.

Sport-hunting of Cheetahs and international trade of their skins is currently legal only in Namibia (where this photograph was taken) and Zimbabwe. There is also substantial illegal trade in live Cheetahs across borders, mainly in southern and north-east Africa.

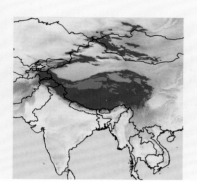

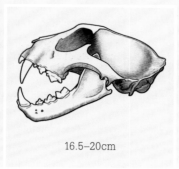

16.5–20cm

● **IUCN RED LIST (2008):** Endangered
Head-body length ♀ 86–117cm, ♂ 104–125cm
Tail 78–105cm
Weight ♀ 21–53kg, ♂ 25–55kg

Snow Leopard

Panthera uncia (Schreber, 1775)

Ounce

Taxonomy and phylogeny

The Snow Leopard was long considered the only member of the genus *Uncia* based largely on its relatively unusual, domed skull (compared to other large cats), but genetic analysis demonstrates that it is closely related to the 'big cats' and is now classified in the genus *Panthera*. It is thought to have arisen from an early branching within the lineage and is most closely related to the Tiger, with a common ancestor more than two million years ago, though the relative position of these two species within *Panthera* is poorly understood. Two Snow Leopard subspecies are sometimes recognised: *P. u. uncia* (central Asia, north-east to Mongolia and Russia) and *P. u. uncioides* (western China and Himalayas), based on superficial morphological differences that are unlikely to be meaningful and await genetic analysis.

Description

The Snow Leopard is the smallest member of the genus *Panthera*, moderately smaller and more lightly built than the Leopard with dense, long fur that gives the impression of a much larger animal. The deep chest and forequarters are muscular, and the legs are relatively short and robust with very large feet. The tubular tail is thickly muscled and proportionally the longest of any felid, 75–90 per cent of the head-body length. It is wrapped around the body to insulate against extreme cold and is used for balance during hunts. The head is small, broad and rounded, with a short muzzle and high domed forehead caused by enlarged nasal cavities believed to assist with breathing at high altitude. The Snow Leopard's legendary fur is extremely dense and long, up to 5cm on the back and sides and as long as 12cm on the underparts in winter. The fur provides excellent insulation; even in low temperatures, the Snow Leopard will rest on its back with the belly exposed to dissipate heat.

The Snow Leopard's background colour is dark cream to smoky grey with yellowish-cream to crisp white underparts. The body is marked with large dark grey or black open blotches, often with fuzzy edges, graduating to more defined, bold open blotches running in rows along the back and tail. There are smaller solid black blotches on the lower legs and small black spots on the head, neck and shoulders. The coat pattern is unique to individuals and useful for identification (for example, from camera-trap photos), though the markings appear less crisp in the long winter coat and individual recognition is more problematic. There is apparently little regional variation in coat colour and markings. There are no records of melanistic or albino Snow Leopards.

Similar species The Snow Leopard is unmistakable. Leopards from central Asia in pale, long winter coats are sometimes mistaken for Snow Leopards (chiefly in fur markets).

Distribution and habitat

The Snow Leopard is restricted to 12 countries in central Asia, from southern Russia south-west to Uzbekistan; south-east through Mongolia to Gansu, Qinghai and Sichuan, China; and across broad arcs along the Himalayas, Tien Shan and Kunlun mountain ranges that cross China to adjacent countries. The species' distribution closely follows the world's highest mountain ranges including the Altai, Tien Shan, Kunluns, Himalayas, Karakoram,

A Snow Leopard with its kill, an adult male Bharal or Blue Sheep in Hemis National Park, India. The Bharal is a 'significantly preferred' prey species, meaning that Snow Leopards hunt for and kill them more often than by chance encounter alone.

The Snow Leopard's long, muscular tail helps counter balance the body for rapid turns during hunts. It helps endow the cat with an astonishing ability to cover steep, treacherous, rocky slopes at high speeds.

Hindu Kush and Pamirs. An estimated 65 per cent of current Snow Leopard range falls inside China. With international borders falling along high mountain ranges, China shares borders and transboundary Snow Leopard populations with all other range countries except Uzbekistan. Anecdotal reports include the northernmost tip of Kachin State, northern Burma as Snow Leopard range but there are no confirmed records from the country.

Snow Leopards are dependent on alpine and subalpine habitats and always occur in close association with very rugged, rocky terrain. In most of the range, they occur in high mountainous habitat characterised by steep cliffs, deep ravines and high ridges. They use open alpine meadows and alpine scrub, and largely avoid barren land, glaciers and habitats lacking cover. In Mongolia and the Tibetan Plateau, China, Snow Leopards occur in areas of open arid steppe and desert with isolated and relatively low, flat mountainous areas; they are recorded moving up to 80km across wide open expanses between massifs. In some of the range, for example parts of the Chinese Tien Shan, Pakistani Hindu Kush and Russian Altai, mountainous habitats include stands of open coniferous forest in which Snow Leopards occur but they typically avoid dense forest. Snow Leopards cope well with deep snow but in the highest regions they usually spend winters at lower elevations, following the altitudinal movements of ungulates seeking shelter and browse.

The Snow Leopard is mostly found at elevations of 3,000–5,500m except in the northern part of the range where they occur much lower, for example at 900–2,400m in southern Mongolia.

Feeding ecology

The Snow Leopard's range closely overlaps the distribution of various species of mountain ungulates that constitute the most important prey. They are capable of taking prey weighing up to at least 120kg and regularly take adult males of their largest ungulate prey. The diet is typically dominated by one or two large ungulate species, which varies depending on the region and its prey composition. Across the range, the most important species are Asiatic Ibex (or Siberian Ibex), Blue Sheep (or Bharal) and Argali. The ranges of these species individually (in the case of Argali) or collectively overlap that of the Snow Leopard's almost exactly. The Himalayan Tahr is also a major prey species where it occurs in the Himalayas from Sikkim to Kashmir. Other ungulate prey includes Markhor, Urial, Himalayan Musk Deer, Roe Deer and rarely gazelles, Eurasian Wild Boar and Asiatic Wild Ass (almost certainly only young individuals of the latter two). Large kills of ungulates are supplemented by small prey, especially during spring and summer when herbivores disperse to high elevations in most of the range. Marmots are the most important smaller prey, but Snow Leopards also

A sub-adult Snow Leopard watches a Eurasian Magpie, northern Pakistan. Birds are a very infrequent occurrence in Snow Leopard diet and never contribute significantly to the species' energetic needs.

opportunistically take small rodents such as voles and hamsters, rabbits, hares, pikas and gamebirds such as Tibetan Snowcock and partridges. Incidental kills of small carnivores are recorded, mainly Red Fox but also martens and weasels. There is a credible record of an 18-month-old Brown Bear killed and largely eaten by two Snow Leopards in Aksu-Zhabagly Nature Reserve, western Tien Shan, Kazakhstan. Snow Leopards kill domestic animals, which often rank second in the diet following wild ungulates. Cattle, domestic yaks, sheep and goats are most often killed, and there are incidental records of domestic camel calves, horse foals and dogs. Depredation can reach significant levels locally and seasonally when wild prey is scarce and when livestock is unattended in montane valleys and meadows. Snow Leopards occasionally enter livestock corrals, sometimes resulting in large losses, for example 82 sheep killed in one night by a Snow Leopard. Importantly, there is evidence that Snow Leopards prefer to prey on wild ungulates even when livestock is abundant. Collared Snow Leopards in the Tost Mountains in the southern Gobi Desert, Mongolia, ate mainly ungulates (comprising 73 per cent of kills compared to 27 per cent made up of livestock) despite livestock being 10 times more

abundant in the area. Snow Leopards do not prey on humans. There are extremely few recorded attacks on people, none fatal and most of them the result of extreme provocation. Two men attacked in 1940 in south-eastern Kazakhstan were seriously injured, but the Snow Leopard was infected with rabies.

The Snow Leopard hunts mainly at dusk through the night into the early morning, though diurnal hunting is more common during winter and where people do not occur. Snow Leopards search for prey by walking along key features of the landscape such as high ridges, animal trails, watercourses and valley bottoms, and resting in wait at strategic sites such as overlooks, terraces and near water sources or salt licks. When prey is located, the Snow Leopard stalks in typical felid fashion, often attempting to reach a position above the quarry before rushing it at close range. The Snow Leopard is superbly sure-footed and is able to pursue prey over extraordinarily steep, rugged terrain. The chase lasts for 200–300m at most, during which the cat attempts to hook the prey with its dew-claw or knock it off its feet (occasionally resulting in prey falling hundreds of feet and being lost). Large prey is usually killed by asphyxiation with a throat bite. There is no information on Snow Leopard hunting success. Unless disturbed, Snow Leopards feed on large kills until finished, for up to a week. They eat carrion, including occasionally from kleptoparasitism; there is one observation of a Snow Leopard displacing four Dholes from a domestic goat they had just killed (Hemis National Park, India). Snow Leopards in turn may lose kills to other carnivores, including Grey Wolves and Brown Bears. On the Tibetan Plateau, China, domestic dogs living in association with Tibetan monasteries harass Snow Leopards and regularly drive them off kills.

Social and spatial behaviour

Snow Leopards display solitary, territorial behaviour including regular marking with urine, faeces and scrapes, but the extent to which ranges are exclusive and defended is poorly known. Before 2009, only 14 Snow Leopards had been radio-collared, in India,

Mongolia and Nepal. Between 2009 and 2013, 19 individuals were fitted with GPS collars in the Tost Mountains, southern Mongolia and three Snow Leopards have been GPS-collared in the Hindu Kush Mountains, Afghanistan (2012). Both studies are expected to produce a more detailed understanding of the species' socio-spatial ecology. Range size is likely to be large given the naturally low prey densities of their habitat. Home range estimates from ground-based radio-telemetry prior to 1998 are likely to be gross underestimates; a Mongolian female's calculated range went from 58km² to at least 1,590km² (and possibly more than 4,500km²) when her VHF radio-collar was replaced with a satellite collar. Range size in the Tost Mountains is 87.2–193.2km² (adult females) and 114.3–394.1km² (adult males) using conservative estimates that assume Snow Leopards closely follow the edges of mountain ranges. This increases to 202.3–548.5km² (females) and 264.9–1,283km² (males) if associated steppe areas are included, but Snow Leopards rarely spend any time in steppe habitat. Snow Leopards are able to cover large distances in rugged terrain, often 10–12km a day and as long as 28km. Snow leopards naturally occur at low densities and are extremely difficult to count. Estimates from camera-trapping include 0.15 Snow Leopards per 100km² (Sarychat, Kyrgyzstan), 1.5–2.3 per100 km² (Tost, Mongolia) and 4.5 per 100km² in prey-rich habitat (Hemis National Park, India).

Reproduction and demography

There is little information from the wild but reproduction in Snow Leopards is almost certainly strongly seasonal, unlike in other *Panthera* cats. Winters are extreme everywhere in Snow Leopard range and even captives show marked seasonality, with births occurring from February to September; 89 per cent of births occur April through June. This is consistent with elevated calling and scent-marking in wild Snow Leopards that peaks between January and March during the presumed mating period. According to a small number of available records,

young cubs in the wild occur from April to July. Oestrus lasts 2–12 days (usually five to eight days) and gestation lasts 90–105 days. Litter size averages two to three cubs, exceptionally up to five. Weaning occurs at two to three months. Age at independence is poorly known; two Mongolian subadults left their mother at 18–24 months, coinciding with the mother's return to oestrus. Sexual maturity occurs at around two years for both sexes in captivity; wild Mongolian females first reproduce at three to four years. A minimum of 21 cubs (and estimated 32 cubs) were born to a population of 12–14 adult Snow Leopards over four years in the Tost Mountains, southern Mongolia.

Large and uncontrolled feral dog populations are a major threat to wildlife populations in parts of Snow Leopard range, particularly where Tibetan Buddhism – which prohibits killing animals, including stray dogs – is prevalent. Dogs kill Snow Leopard prey, appropriate their kills and occasionally kill cubs.

Snow Leopards are not aggressive and, sadly, are easily subdued by people. Both cubs and adults are sometimes captured alive by local people for sale into the illegal pet or parts trade.

Mortality Based on four years of camera-trap records, annual mortality of Snow Leopards in a relatively well-protected population (Tost Mountains, Mongolia) has been estimated at 17 per cent (adults) and 23 per cent (subadults). Causes of natural mortality are poorly known. Starvation is likely to be an important seasonal factor especially for young animals given the extremes in weather in Snow Leopard habitat. Predation is probably very rare but may occur, especially on cubs; potential predators include the Grey Wolf and Brown Bear. Most recorded mortality is by people.

Lifespan Unknown in the wild; up to 20 years in captivity.

STATUS AND THREATS

Snow Leopards are extremely challenging to find and survey, and their status is poorly known across the majority of the range. The total population is crudely estimated at 4,000–7,000, with between 50 and 62 per cent of all wild Snow Leopards estimated to occur in China. The presence of Snow Leopards is regarded as definitive or probable in only 41 per cent of the estimated 2.9 million km² current range. With such poor knowledge, it is difficult to estimate the areas from which Snow Leopards have been extirpated but it is thought to be far less than for most large felids, perhaps as little as 5–10 per cent of historical range. Known areas of range loss include parts of central and northern Mongolia and southern Siberia, though Snow Leopard numbers may never have been high in any of these regions. Snow Leopard numbers are thought to have declined by as much as 40 per cent in parts of the former Soviet Union during the 1990s, when associated economic collapse led to widespread poaching of Snow Leopards and their prey. Those declines have been halted and in some cases reversed due in part to improved economic conditions and intensive conservation efforts. Snow Leopard numbers are thought to be increasing in some parts of the Himalayas, and they have reappeared in such places as the Mount Everest region where they had not been seen in nearly three decades. Over much of their range, populations may be stable or slightly declining, although solid data is lacking in most cases.

Snow Leopards are somewhat insulated from human activities given they live in such remote, inhospitable areas but they are naturally rare, and human populations and their livestock are increasing in Snow Leopard habitat. The main threat to Snow Leopards relates to the depletion of wild ungulates that are widely hunted by people. This is compounded by the presence of livestock that potentially displaces wild ungulates and is occasionally preyed upon by Snow Leopards. In retaliation, Snow Leopards are widely persecuted for livestock depredation. Recent and ongoing poisoning campaigns of marmots and pikas in China and Mongolia may exacerbate the threat of prey depletion. International trade in Snow Leopard fur once averaged around 1,000 skins a year. Fortunately this is now illegal although furs and especially body parts continue to be illegally traded, driven by demand mainly from China, Indochina and eastern Europe. Trade in Snow Leopard pelts and bones in China was formerly a local problem restricted to Snow Leopard range provinces but is reported to be emerging in the wealthier coastal cities. The demand for luxury rugs and taxidermy, especially from China and eastern Europe, also appears on the increase. Skins and carcasses from animals killed in retaliation by herders are often sold into the trade. Emerging threats for the species include rapid expansion of mining, road and rail construction, and intensified hydroelectricity development in much of central Asia and the Himalayas. Over the longer term, climate change is expected to impact Snow Leopard habitats, and likely lead to further intensification of human presence and use.

CITES Appendix I. Red List: Endangered. Population trend: Decreasing.

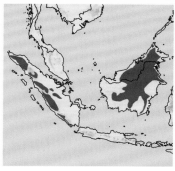

Sunda Clouded Leopard

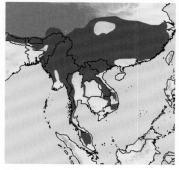

Indochinese Clouded Leopard

IUCN RED LIST (2008): Vulnerable

Head-body length ♀ 68.6–94cm, ♂ 81.3–108cm
Tail 60–92cm
Weight ♀ 10–11.5kg, ♂ 17.7–25kg

Sunda Clouded Leopard

Neofelis diardi (G. Cuvier, 1823)

Indochinese Clouded Leopard

Neofelis nebulosa (Griffith, 1821)

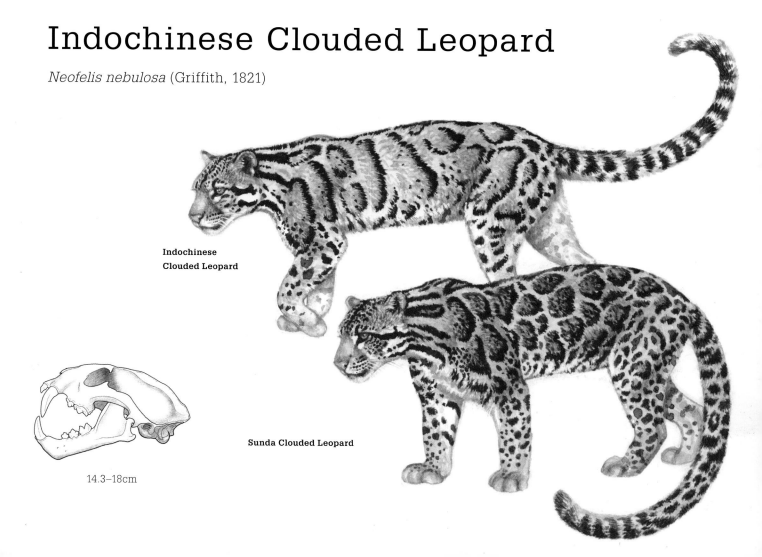

Indochinese
Clouded Leopard

Sunda Clouded Leopard

14.3–18cm

Taxonomy and phylogeny

Until 2006, the Clouded Leopard was considered a single species classified in its own genus, *Neofelis*. Genetic analyses published in 2006–2007 strongly suggest that island populations on Borneo and Sumatra have been reproductively isolated from mainland populations for 1.4–2.9 million years, well within the range of species-level divisions between other large cats. These molecular distinctions are supported by differences in the pelage, and in measurements of the skulls and teeth. Taken collectively, these differences are now considered sufficient to recognise clouded leopards as two species: *N. diardi* (Sunda Clouded Leopard or Sundaland Clouded Leopard; Borneo and Sumatra) and *N. nebulosa* (Indochinese Clouded Leopard or simply Clouded Leopard; mainland Asia). Each

species is provisionally divided into two subspecies. There is limited genetic data supporting the subspecific separation of the Borneo population of the Sunda Clouded Leopard (*N. diardi borneensis*) from that on Sumatra (*N. diardi sumatrensis*). The mainland population of the Indochinese Clouded Leopard is historically split into eastern (*N. nebulosa nebulosa*) and western (*N. nebulosa macrosceloides*) subspecies that have minor morphological differences yet no significant pelage or genetic differences. A third putative subspecies on Taiwan, *N. nebulosa brachyura*, is likely extinct. Clouded Leopards are closely related to the 'big cats' and are grouped with them in the *Panthera* lineage. *Neofelis* diverged very early from the line that led to the genus *Panthera*, with a common ancestor estimated around 6.4 million years ago.

Below: **It is unclear if the two Clouded Leopard species can interbreed. It is hypothetically possible given their relatively recent divergence, although there is no natural range overlap except perhaps in Peninsular Malaysia (where genetic analysis is yet to confirm if one or both clouded leopard species are present).**

Description

Clouded Leopards of both species are very similar in general appearance and size; based on very few samples, there is limited evidence Sunda Clouded Leopard males are larger than mainland males. Clouded Leopards have a long body, relatively short robust legs with large feet and a very long tail. The head is long and heavily built, similar in overall shape and proportions to the *Panthera* cats. Clouded Leopards have an exceptionally large gape (almost 90° compared to 65° for the Puma) and elongated canine teeth measuring up to 4cm, relatively the longest of any living felid. This is similar to extinct sabre-toothed cats (to which all modern cats including Clouded Leopards are not closely related), the reasons for which are unclear. Sunda Clouded Leopards of both sexes have the longest canines by a

small margin. Pelage differences between the two Clouded Leopards are distinct. Sunda Clouded Leopards are overall darker with grey to greyish-yellow background fur and relatively small, irregular blotches with thick, black margins and usually small black spots within each blotch. The lower legs are marked with solid, black blotches that are closely clustered. Indochinese Clouded Leopards are generally paler and brighter with buff to rich tawny background colour, and very large blotches with narrower black margins and few or no spots within the blotches. Solid black blotches on the lower legs tend to be more widely spaced than in Sunda cats. Melanism is reported anecdotally, though there is no physical evidence.

Similar species The similar Marbled Cat is sympatric with both clouded leopard species.

Opposite and below:
A Sunda Clouded Leopard (opposite) and Indochinese Clouded Leopard (below, C), showing the distinct differences in pelage. There are also significant differences between the two species in skull morphology in which adaptations similar to those in extinct sabre-tooths are more advanced in the Sunda Clouded Leopard.

Clouded Leopards are much larger with a distinctly longer, heavier head typical of a 'big cat' compared to the small rounded head of the Marbled Cat. Both Clouded Leopard species have larger, more discrete blotches with distinct edges compared to the more diffuse markings of the Marbled Cat.

Distribution and habitat

The Sunda Clouded Leopard is endemic to Borneo and Sumatra. On Borneo it occurs in Kalimantan (Indonesia), Sabah and Sarawak (Malaysia) and Brunei, potentially on approximately 50 per cent of the island where there is still good forest cover. It is absent (or likely so) in large deforested areas mainly in the south-east and west, and along a wide coastal band. On Sumatra, it occurs largely along the Bukit Barisan mountain range running the length of the island. It is unknown if Sunda Clouded Leopards occur on the offshore Batu Islands close to Sumatra. Clouded Leopards occurred on Java perhaps as

recently as the Holocene (starting 11,700 years before the present) but not in modern times, and they have never occurred on Bali. The Indochinese Clouded Leopard occurs in central Nepal, Bhutan, north-east India and eastern Bangladesh (marginally), across China south of the Yangtze River, and patchily throughout Indochina including the Malay Peninsula (which is usually regarded as part of the Sundaland region, south of the Isthmus of Kra; however, based on their pelage, the Clouded Leopards on the peninsula are considered *N. nebulosa*). The species is considered extinct on Taiwan.

Clouded Leopards are closely associated with dense habitat and are considered forest-dependent. They occur in all kinds of dense moist and dry forest, peat-swamp forest, dry woodlands and mangroves from sea level to 1,500m (Sarawak) and 3,000m in the Himalayan foothills. They use grassland patches in forest-grassland mosaics, and a radio-collared cat in Nepal rested in dense, 4–6m-tall grass on *terai*

Below: **During observed attacks on Proboscis Monkey troops by Sunda Clouded Leopards, adult male monkeys attempt to defend young monkeys. In one incident 10m above ground, a male Probsocis drove off a Clouded Leopard that had captured a 10-month-old infant but not before it was killed.**

floodplain grasslands. Clouded Leopards appear to tolerate some habitat modification. They may be relatively common in secondary and selectively logged forest but their ability to use forestry areas seems to decrease beyond a threshold of fairly light logging intensity. Clouded Leopards occasionally use oil-palm plantations but all camera-trap evidence indicates they largely use only the periphery of plantations. A radio-collared male in Borneo crossed oil-palm plantations under 1km-wide between forest patches and made one foray of around 2.5km through an oil palm-scrub mosaic; he did not linger in oil palm habitat and always moved rapidly through it.

Feeding ecology

Clouded Leopard ecology is poorly known. Based on mostly anecdotal and incidental records, they are known to prey on a wide variety of small and medium-sized vertebrates including both terrestrial and arboreal species, and diurnal and nocturnal species. Primates and small ungulates possibly constitute the staple prey. However, as in most felids, the diet is presumably flexible and varies regionally depending on the composition of the prey community and the presence of other carnivores. Clouded Leopards are the largest felid with few competitors on Borneo whereas Tigers, Leopards and Dholes co-occur on mainland Asia and on Sumatra (excluding Leopards). It is unknown if or how these species affect Clouded Leopard feeding patterns.

Clouded Leopards are recorded killing a range of primates, from small, nocturnal slow lorises to large, diurnal adult male Proboscis Monkeys. Anecdotes of predation on Orang-utans have not been confirmed. They are capable of taking ungulates at least up to their own size, with reliable records of Barking Deer, Hog Deer (including adults) and Bearded Pig (probably young animals). Smaller prey includes oriental chevrotains, Malayan Pangolins, brush-tailed porcupines, small rodents including Indochinese Ground Squirrels, Binturongs, Common Palm Civets, and a variety of birds. Local people in Sabah and Sarawak report that fish are sometimes eaten. Clouded Leopards occasionally kill poultry; a subadult male Clouded Leopard in Nepal was trapped in a chicken coop (and subsequently released with a radio-collar). Attacks on livestock are rare. One Sunda Clouded Leopard was shot after reportedly killing goats in an enclave village surrounded by forest.

The Clouded Leopard's morphology suggests a high degree of arborealism, with short robust legs, broad feet and a long tail. There are numerous accounts of hunting above ground, including four published observations of Clouded Leopards attacking Proboscis Monkeys in trees, one resulting in the capture of a juvenile monkey 7m above ground. However, based on a small number of data points, radio-tracked individuals travelled and hunted mostly on the ground; Clouded Leopards probably search for prey mainly from the ground and readily pursue arboreal species into trees as they are located. They hunt primarily at night with

A Clouded Leopard attacks a Hog Deer. With so little known about the Clouded Leopard's killing technique, it is still unclear whether the species routinely uses the deep neck bite observed on a handful of carcasses.

The Clouded Leopard's blade-like upper canines and wide gape are among many features it shares with extinct sabretooths including a lowered jaw joint and a reinforced mandible that is resistant to bending.

crepuscular activity peaks, although two collared individuals in Thailand were often active in the morning until midday. One of these radio-collared cats, an adult male, hunted Hog Deer and Barking Deer as they bedded down en masse on open grasslands at dusk. This Clouded Leopard rested at the forest edge before moving out onto the grasslands at night to target the deer. One of its kills, an adult male Hog Deer, was killed with a 3cm deep

bite through the spine above the shoulders. The same technique was reported by local people in Sabah and Sarawak who found Clouded Leopard kills of deer and pigs. Similarly, two young Proboscis Monkeys were both attacked with bites to the back of the head and neck. This is a very unusual killing technique compared to other felids, which typically kill large prey and especially ungulates from asphyxiation by biting the throat, and it may be related to the Clouded Leopard's unique dentition. It is unknown if Clouded Leopards scavenge.

Social and spatial behaviour

Only 12 Clouded Leopards have ever been radio-collared: five in Borneo and seven in Nepal and Thailand. The latter two studies provided a limited amount of data over a short period; the former study is ongoing with GPS telemetry that is expected to provide greater detail. Based on the little available information, Clouded Leopards are thought to follow a typical felid pattern of overlapping territories with exclusive core areas. Limited data suggest the sexes have similar territory sizes although it is likely that males have larger ranges than females and overlap multiple female territories; this awaits more research. The published range sizes are 16.1–40km^2 (two subadult females and two adult females) and 35.5–43.5km^2 (one subadult male and two adult males). Density estimates appear to be quite low compared to other cats, especially on Borneo despite the lack of competing large carnivores. Density estimates for Sunda Clouded Leopards include 0.84–1.04 per 100 km^2 in degraded secondary lowland forest (Tangkulap-Pinangah and Segaliud Lokan Forest Reserves, Sabah); 1.29 per 100km^2 (Tesso Nilo-Bukit Tigapuluh Conservation Landscape, Sumatra); 1.9 per 100km^2 for primary forest (Maliau Basin Conservation Area, Sabah) which drops to 0.8 per 100km^2 when adjacent logged forest is included in the estimate; 1.76 per 100km^2 in well-protected primary forest (Danum Valley Conservation Area, Sabah); and 2.55 per 100km^2 in secondary forest undergoing rehabilitation (Ulu

Segama Forest Reserve, Sabah). The only rigorous density estimate for Indochinese Clouded Leopard is 4.7 per 100km² (Manas National Park, India).

Reproduction and demography

This is virtually unknown from the wild. Most information comes from captive animals (all of them Indochinese Clouded Leopards) and is based on few records. Breeding occurs year-round in captivity and is likely to be aseasonal in the wild. Unusually, males quite frequently kill females in captivity (with a similar spinal nape bite as for large prey), the reasons for which are unknown but assumed to be related to a captive setting; it is extremely unlikely that this behaviour is common for wild individuals. Oestrus lasts about one week and gestation lasts 85–95 days (rarely to 109 days). Litters average two to three kittens, exceptionally reaching five. Weaning begins around 7–10 weeks. Sexual maturity is 20–30 months.

Mortality People are responsible for most deaths in studied areas, otherwise mortality is unknown. Large carnivores are potential predators, though there are no records. Clouded Leopards take refuge in trees when pursued by domestic dogs.

Lifespan Unknown in the wild; up to 17 years in captivity.

The Clouded Leopard is half the size of the next smallest species of the *Panthera* lineage, but its big cat ancestry is clear in the elongated, heavily built head and jaws with a relatively smaller braincase compared to small cats.

STATUS AND THREATS

Clouded Leopards are relatively widespread but their status is poorly known in most of the range. Although they are often regarded as more resilient than larger felids, they do not appear to attain high densities anywhere and they are closely associated with forested habitats which are undergoing extremely rapid conversion by people in much of the range. South-east Asia has the world's fastest deforestation rate due to logging and conversion for settlement and agriculture, including plantations especially of oil palm and rubber. Forest loss is the main factor driving declines and has contributed to extirpation of the Sunda Clouded Leopard from an estimated 50 per cent of Borneo and around two-thirds of Sumatra. The Indochinese Clouded Leopard has undergone similar range loss and decline, particularly in China and in Indochina where the distribution is very patchy, and there are few recent records for most of Cambodia, China, Laos and Vietnam. Clouded Leopard skins, bones and meat have commercial value, and are illegally traded in wildlife markets; for example, 13 market surveys between 2001 and 2010 at two border towns in Burma recorded parts of at least 149 individuals. Clouded Leopards are vulnerable to snares and are hunted opportunistically in much of their range including in protected areas; a minimum number of seven were killed inside Kerinci Seblat National Park (Sumatra) in 2000–2001.

Both species: CITES Appendix I. Red List: Vulnerable. Population trend: Decreasing.

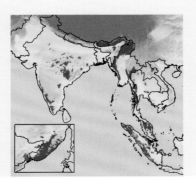

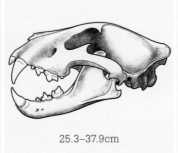

25.3–37.9cm

IUCN RED LIST (2015):
Endangered (global)
Critically Endangered
(Malayan and Sumatran Tiger)

Head-body length ♀ 146–177cm,
♂ 189–300cm
Tail 72–109cm
Weight ♀ 75–177kg, ♂ 100–261kg

Tiger

Panthera tigris (Linnaeus, 1758)

Bengal Tiger

White form

Sumatran Tiger

Amur Tiger

Taxonomy and phylogeny

The Tiger is one of the 'big cats' in the genus *Panthera* and thought to be most closely related to the Snow Leopard with a common ancestor more than two million years ago, though the relative position of these two species within *Panthera* is uncertain. The Tiger is traditionally classified into eight subspecies, three of which are extinct (Javan *P. t. sondaica*, Bali *P. t. balica* and Caspian *P. t. virgata*). A fourth, the South China Tiger (*P. t. amoyensis*), is almost certainly extinct in the wild with no unequivocal evidence since the 1970s. Approximately 60–70 are maintained in captivity in China, though most have evidence of hybridisation with other subspecies. Recent molecular analyses support the division of extant tiger populations into four classically described subspecies with possibly a fifth: Amur Tiger *P. t. altaica* (Russian Far East,

marginally east China); Indochinese Tiger *P. t. corbetti* (Indochina); Bengal Tiger *P. t. tigris* (Indian subcontinent); Sumatran Tiger *P. t. sumatrae* (Sumatra) and, controversially, Malayan Tiger *P. t. jacksoni* (Malayan Peninsula). Importantly, the molecular differences between Tiger subspecies are modest; they intergrade at the putative boundaries and arose very recently, likely well within the last 100,000 years. The Sumatran Tiger is clearly sufficiently genetically isolated to be accorded its own subspecific status. The support for the Malayan Tiger as a distinct subspecies is least convincing. It has slight genetic differences from the Indochinese Tiger and does not differ in morphology (cranial and pelage characteristics). Similarly, samples from the now extinct Caspian Tiger show very few genetic differences to the Amur Tiger. Continental Tiger populations were probably largely continuous until

Young Tigers at play in Tadoba-Andhari Tiger Reserve in tropical dry forest typical of central India and considered to be excellent habitat for the species.

A male Sumatran Tiger showing the species' distinctive facial ruff which tends to be most prominent in this subspecies and the Amur Tiger. The Sumatran Tiger is the last island subspecies remaining, and is genetically and morphologically distinct from all mainland populations (C).

sport-hunted animals (from the early 20th century) are often inflated. The Tiger is a massive, powerfully built cat with a deep chest, muscular forequarters and heavily built limbs. Body size varies across the range, roughly decreasing in a cline from north to south, and correlated with prey availability. Tigers from the Indian subcontinent and Russian Far East are largest, and Sumatran Tigers (and extinct island populations from Java and Bali) are the smallest. Wild male Sumatran Tigers weigh up to 140kg while the largest wild male on record, from Nepal, weighed 261kg (up to 325kg is recorded from captivity).

The Tiger's ground colour varies from pale yellow to rich red with white or cream underparts. Tigers are generally darker and more richly striped in tropical South-east Asia and paler, more lightly striped in temperate areas. The fur on the cheeks forms a long facial ruff that is most developed in males of Amur and Sumatran Tigers. Amur Tigers (and the now extinct Caspian form) develop a long, dense winter coat which has a washed-out appearance compared to summer pelage. White Tigers are not albino and arise from a recessive mutation that produces blue eyes and chocolate-coloured stripes on a white background. There is only one record from the wild (Madhya Pradesh, India) since 1951, a male cub from which all captive white Tigers are descended (and are extremely inbred as a result). An intermediate form called 'golden tabby' or 'strawberry' is known only from captivity. Complete melanism is unknown, though individuals occasionally appear with pseudomelanism (or abundism), in which the stripes are very wide and merged, producing an almost entirely black appearance.

Similar species The Tiger is the only striped cat and is unmistakable. Local nomenclature may cause confusion as variants of the name for Tiger are often employed for other cats.

relatively recently, and some authorities argue that all mainland populations should be treated as the same, single subspecies with distinct but slight differences between populations that should be treated as 'evolutionary significant units' rather than subspecies. The oldest Tiger fossils, about two million years old, are from northern China and Java.

Description

The Tiger is the world's largest cat by a small margin. The Lion is comparable in all measurements and Lions have longer skulls on average, but the largest Tigers have slightly longer and heavier bodies than the largest Lions. It is important to note that there are few reliable body measurements of wild Tigers, and that figures from captivity and of

Distribution and habitat

The Tiger is now restricted to, at most 10 per cent of its historical range with an extremely patchy distribution largely restricted to isolated populations in forest fragments. In South Asia, it occurs mainly in protected areas in India (chiefly south-western, central and north-eastern India); in a narrow band along the southern lowlands and slopes of the Himalayas from Uttarakhand, India, through Nepal and Bhutan to Arunachal Pradesh, India; and a population in the Sundarbans of Bangladesh and India. The only definite populations remaining in Indochina occur in western Thailand bordering Burma (Huai Kha Khaeng/Thung Yai Naresuan, Kaeng Krachan/Kui Buri) and Peninsular Malaysia (Endau Rompin, Taman Negara National Park, Belum-Temengor Forest). Tigers may still occur more widely in extensive forest tracts along the Thai-Burma border, and in western and northern Burma, though there are few recent records and no evidence of reproduction. In Sumatra, Tigers occur patchily along the Bukit Barisan mountain range running the length of the island and in one population in central Sumatra. The largest contiguous Tiger population is in the Russian Far East and marginally into adjacent areas of eastern China; in contrast to the rest of the range, most of this population occurs outside protected areas. The Tiger is now functionally or entirely extinct in China (except for the Amur population), Cambodia, Lao PDR, North Korea and Vietnam. It is extinct in Bali (1940s), central Asia (1968) and Java (1980s). Tigers are naturally absent from Sri Lanka and Borneo.

Tigers occur in a variety of tropical, subtropical and temperate forests, forest-grassland mosaics and associated dense cover such as *terai* (dense floodplain grasslands), thickets, scrub and marshes. They reach highest densities on the Indian subcontinent in dry and mesic forests, and *terai*. The Sundarbans population (Bangladesh–India) lives in low-lying freshwater swamp-forest and salt-tolerant mangrove forests inundated by tidal seawater. The Amur Tiger inhabits mountainous terrain with temperate forest of Korean Pine, birch, fir, oak and spruce with deep winter snowfall and temperatures reaching -40°C. They traverse but do not permanently occupy human-modified habitats such as agriculture, palm plantations and monocultures. Tigers occur from sea level typically to 2,000m, and are recorded from montane forest with deep snow up to 4,201m in the Himalayas.

A rare photograph of a wild female Amur Tiger, taken in Lazovskiy Zapovednik (strictly protected area) on the Russian coast of the Sea of Japan at the southern tip of the subspecies' present range. Genetic analysis of extinct Caspian Tiger samples from central Asia show minimal differences from Amur Tigers and they probably comprised a single population perhaps as recently as the last 200 years.

194 Wild Cats of the World

Feeding ecology

The Tiger is an immensely powerful predator adapted to overpower prey as large as or larger than itself. A healthy adult Tiger can kill almost anything it encounters with the exception of adult rhinos and Asian Elephants, but the diet is dominated by various medium-sized and large deer species and wild boar. Tigers typically focus on two to five species of locally common ungulates weighing 60–250kg, particularly Sambar, Red Deer, Chital, Hog Deer, muntjacs and Wild Boar. They are capable of killing adult Gaur and Asiatic Water Buffalo exceeding 1,000kg, though most kills of these species are juveniles and subadults. Regional dietary patterns are best known from the Indian subcontinent and the Russian Far East where Tigers are best studied. The staple of Tigers in India and Nepal is Chital, Sambar and Wild Boar, as well as Barking Deer (Indian Muntjac) and Hog Deer

Tigers opportunistically kill primates but they rarely contribute significantly to intake. Grey Langurs are the primate most commonly killed by Tigers but even where they are super-abundant rarely comprise more than 2–3 per cent of Tiger diet by mass.

depending on local abundance. Barasingha, Chousingha, Chinkara Gazelle, Blackbuck, Nilgai, Himalayan Goral and Nilgiri Tahr are also recorded. Tigers in Nagarahole-Bandipur reserves in India's Western Ghats prey heavily on Gaur as well as Sambar. Amur Tigers in the Russian Far East rely chiefly on Red Deer and Wild Boar (84 per cent of 552 documented kills of wild prey; 64 per cent of 729 kills, including domestic animals) with Siberian Roe Deer and Sika Deer common in the diet in some areas. Siberian Musk Deer, Moose and Long-tailed Goral are recorded incidentally. Tiger diet is least known from South-east Asia but based on a small number of records, Barking Deer, Wild Boar and Sambar are most important in Thailand and Way Kambas National Park, Sumatra. Tigers in Way Kambas National Park also frequently killed Southern Pig-tailed Macaques. The diet in South-east Asian lowland forest is thought to include a greater

killed in winter dens, Russian Far East). Carnivore carcasses are often abandoned without having been eaten, though bears are usually partially or entirely consumed. Cannibalism occurs rarely, generally of cubs killed by infanticidal males and occasionally of adults killed in territorial clashes. Tigers prey on livestock, mainly when unattended in forest. Together with wild Sambars, domestic cattle, yaks and horses comprised 93.3 per cent of Tiger diet during one study in mountainous habitat with very low densities of wild prey where livestock roams freely (Jigme Singye Wangchuck National Park, Bhutan). Tigers were not recorded killing domestic dogs during this study, but Amur tigers regularly kill dogs (87 of 177 recorded kills of domestic animals), particularly when dogs are accompanying hunters in the forest and during severe winters that force Tigers into villages seeking prey. Tigers probably kill more people than any other large carnivore, in part because human population density in Asia is so high and people utilise Tiger habitat intensively. Habitual 'man-eaters' that focus on humans as prey are very rare.

Hunting is mainly nocturno-crepuscular and terrestrial. Tigers are recorded climbing to 7.5m but they are too heavy to hunt arboreally except for occasional cases of prey snatched from lower branches. Tigers search for prey as they walk, readily using forest roads, logging tracks, animal trails and watercourses. Depending on prey availability, Tigers cover large distances while hunting: 3–10km in prey-abundant areas to more than 20km a night where prey occurs in low densities (Russian Far East). They intersperse mobile hunting with waiting or resting where prey congregates, for example near water sources or salt licks and at the edges of meadows and open areas. The Tiger stalks prey in typical felid fashion before an explosive rush, usually from within 25m of the target. Tigers usually abandon pursuit if a kill is not made after 150–200m; most successful hunts are shorter. Tigers kill most prey by asphyxiation with a throat bite or by covering the nostrils and mouth

The Wild Boar is the second most important prey species (after Red Deer) to Amur Tigers, comprising around a quarter of the diet by mass. It is assumed to also be important to Tigers in South-east Asia where large ungulates are uncommon, but few data exist (C).

diversity of smaller prey given the low productivity of the habitat. Banteng and Malayan Tapir are recorded as rare prey in Malaysia and Thailand.

Smaller prey taken relatively often by Tigers includes primates (especially Grey Langurs; Rhesus Monkeys are consumed quite often in the Sundarbans), Indian Crested Porcupines, hares, small carnivores (including records of Yellow-throated Marten, Greater Hog Badger, Eurasian Badger, Golden Jackal, Red Fox, Raccoon Dog, civets and mongooses) and birds such as peafowl. Herptiles, fish and crabs are eaten, though the contribution to intake is trivial. Tigers, including females, are recorded killing very large Mugger Crocodiles up to 4m long and poisonous snakes; an adult male Tiger found dead had eaten a King Cobra and an Indian Cobra (Sundarbans Tiger Reserve, India). Tigers kill other felids and large carnivores including Jungle Cats, Fishing Cats (Sundarbans Tiger Reserve, India), Asiatic Golden Cats (Chitwan National Park, Nepal), Eurasian Lynx (Russian Far East), Leopards, Dholes, Grey Wolves, Asiatic Black Bears, Sloth Bears and Brown Bears (including adults with a clamping muzzle bite, the latter used most often for very large prey such as adult Gaur. They kill very large crocodiles by biting the spine where it meets the base of the skull. Small prey is bitten at the nape or skull.

Estimates of hunting success by Tigers are poorly known. Based on snow-tracking (which provides limited but accurate data), Amur Tigers hunting Red Deer and Wild Boar in winter were successful 38 per cent and 54 per cent of the time respectively. Success outside winter and elsewhere in the range is expected to be significantly lower. GPS-collared Amur Tigers made a kill (mainly of ungulates) on average every 6.5 days, eating close to 9kg of meat per day. Tigers killed slightly more often, killed more large prey and ate slightly more during winter compared to summer. Unless disturbed, Tigers remain with their kills until finished, spending up to five to six days feeding from large carcasses. Tigers scavenge and readily appropriate the kills of other carnivores including from other Tigers, Leopards, Dholes and Golden Jackals.

Social and spatial behaviour

The Tiger is solitary and generally territorial. Adults socialise chiefly for breeding, but males in high-density populations frequently spend time with familiar females and cubs, including sharing kills. Males are very tolerant of cubs belonging to known females (presumably as the males are likely to be the cubs' sires). Males have large ranges that overlap the smaller ranges of one or more females (for example, two to seven females, in Chitwan National Park, Nepal). Adults establish exclusive territories where possible but complete exclusivity is rare except in small core areas within the range. Territorial overlap is least in high-density populations with abundant prey and small home ranges (for example, in India and Nepal), and greatest where prey is dispersed and ranges are very large (for example, in Russia). Both sexes demarcate territory and advertise their presence by depositing scent-marks via cheek-rubbing and spraying vegetation, and by depositing

scats and scraping the ground with the hindfeet. Territorial fights are rare but are sometimes fatal when they occur, more often in males and more likely to occur during social disruption, such as when a resident dies or an immigrant male moves in. Adult females generally live out their lives in one area while male territorial tenure is typically shorter, on average 2.8 years (ranging from seven months to 6.5 years) in Chitwan National Park, Nepal.

Recorded territory size varies from 10km² (female, Chitwan National Park, Nepal) to more than 1,000km² (male, Russian Far East). Home range size by radio-telemetry has been established only for populations in India, Nepal and Russia. Range size for Tigers in very productive *terai* and forest in Nepal and India is 10–51km² (females) and 24–243km² (males), compared to 224–414km² (females) and 800–1,000km² (males) in Russia. Densities of Tiger populations, even in high-quality habitat, are often depressed due to human hunting of prey (even if Tigers are not hunted). In lowland tropical forest at sites where poaching is prevalent, density estimates include 0.2–2.6 tigers per 100km² (Malaysia, Burma, Sumatra and formerly Laos). In relatively well- to very well-protected lowland forest, density increases to 3.5 Tigers per 100km² (Huai Kha Khaeng, Thailand) and as high as six Tigers per 100km² (southern Tambling Wildlife Nature Conservation, Sumatra). Amur Tigers in temperate forest in the Russian Far East occur at densities of 0.3–1 Tigers per 100km², depending on the level of protection. Tigers reach their highest densities in very productive habitats under strong protection, 8.5–16.8 per 100km² (deciduous forests, alluvial floodplains and *terai*, India).

Reproduction and demography

Tigers are largely aseasonal in their tropical and subtropical ranges, but Amur Tigers show greater seasonal patterns. More than 50 per cent of Amur Tiger cubs are born in late summer (August–October) and winter births are rare. Oestrus lasts two to five days and gestation lasts 95–107 days, averaging

103–105. Litter size is two to five cubs, averaging 2.3–3 (for populations in India, Nepal and Russia). Weaning is around three to five months. Inter-litter interval ranges from 21.6 to 33 months. Cubs reach independence at 17–24 months. Female offspring usually inherit part of their mother's range or settle close by whereas males disperse more widely. Dispersal distances at Chitwan average 9.7km for females (maximum 33km) and 33km for males (maximum 65km). Based on the genetic relatedness of Tigers in Pench Tiger Reserve, India, most females settle within their natal range, with a maximum dispersal distance of 26km; males related to resident females did not occur within 26km, indicating they dispersed further away. Dispersal distances are large for Amur Tigers with some male dispersers reported hundreds of kilometres outside of known Tiger range.

Sexually maturity is 2.5–3 years for both sexes, but breeding in the wild is typically later, ranging from 3.4 to 4.5 years for females, and 3.4 years at the earliest for males (averaging 4.8 years, Chitwan). Females in Chitwan during a period of strong protection and population stability produced litters on average for 6.1 years with a maximum of 12.5 years. Over the course of their lifetime, these females produced on average a total of 4.5 cubs that survived to disperse and only two that survived to breed. Reproduction by females is possible until at least age 15.5.

Mortality 34 per cent (Chitwan National Park) to 41–47 per cent (Russian Far East) of cubs die in the first year, most related to anthropogenic causes and infanticide. Females sometimes defend cubs against infanticidal males. There are two credible records of a female Tiger killing an adult male Tiger while

An adult female (right) Indian Tiger repels the advances of a male, showing the species' marked sexual dimorphism. Adult males exceptionally weigh up to twice as much as females in the same population.

A Tigress cools off with her two-month-old cubs in Ranthambore National Park, India. Tiger reproductive ecology is relatively well known only from a few sites in India, Nepal and the Russian Far East.

defending cubs, apparently surprising the male in both cases. Estimates of annual adult mortality include 23 per cent (both sexes combined, Nagarahole National Park, India) and 19 per cent (females, Russia) to 37 per cent (males, Russia). Humans are the main cause of death for most populations, though natural causes including disease and intraspecific fights can be important. Adult Tigers have few natural predators. There are rare records of large packs of Dholes killing Tigers, perhaps unhealthy or injured individuals; in one eyewitness account, 22 Dholes killed a male Tiger in a prolonged clash in which at least 12 of the dogs were killed. An adult female Tiger was killed by a 4m Saltwater Crocodile in the Indian Sundarbans, apparently while swimming a river. Tigers occasionally die from injuries inflicted by dangerous prey, including verified cases by Water Buffalo, Gaur and Wild Boar. Accidents (for example, a male fell through a frozen river in Russia) are uncommon. Canine distemper has been confirmed in Tigers from Russia and India, though the potential impact on population dynamics is still unclear.

STATUS AND THREATS

The Tiger is the most endangered large cat, having suffered a calamitous decline in the 20th century that continues in much of its range today. The Tiger once ranged widely across Asia, from eastern Turkey through central Asia, and in a massive unbroken swathe from the Afghanistan-Pakistan border through South and South-east Asia to Bali in the east, and the Russian Far East. Since the 1940s, Tigers have disappeared from south-western and central Asia, from Bali and Java, and from most of their range in South-east and eastern Asia. Tigers now occur with certainty in only 4.2 per cent of their historical range, and may occur in a further 5.9 per cent of historical range (that is, where suitable habitat occurs but there is no recent information). There are known breeding populations only in eight range countries (Bangladesh, Bhutan, India, Sumatra/Indonesia, Malaysia, Nepal, Thailand and Russia). The species is likely extinct or at best limited to a few individuals with no recent evidence of breeding in Cambodia, Lao PDR and Vietnam. It is extinct in China except for around a dozen individuals in a transboundary population with Russia. An estimated 70 per cent of all remaining wild Tigers (and probably close to 100 per cent of breeding females) occur in 42 populations, each centred on a protected area that collectively comprise around 100,000km^2, which is less than 0.5 per cent of the historical range. Most of these populations are in India (18), Sumatra (eight) and the Russian Far East (six). Combined with loss of habitat to forestry, commercial palm plantations and agriculture, Tigers are particularly threatened by intense illegal hunting to supply the traditional Chinese medicinal trade. This is compounded by widespread hunting of their prey to feed a massive demand for bushmeat especially in South-east Asia. Given a respite from human hunting, Tiger populations recover quickly; unfortunately, there are now few areas where this is taking place.

CITES Appendix I. Red List: Endangered (global), Critically Endangered (China, Russia, Sumatra).

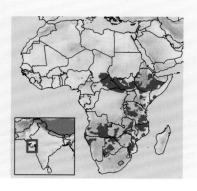

26.7–42cm

IUCN RED LIST (2015):
- Vulnerable (global)
- Endangered (Asia)
- Critically Endangered (West Africa)

Head-body length ♀ 158–192cm,
♂ 172–250cm
Tail 60–100cm
Weight ♀ 110–168kg, ♂ 150–272k

Lion

Panthera leo (Linnaeus, 1758)

Taxonomy and phylogeny

The Lion is one of the 'big cats' in the genus *Panthera* and is most closely related to the Leopard and somewhat more distantly to the Jaguar. The oldest Lion fossils are 3–3.5 million years old from East Africa.

More than 20 subspecies of Lions have been described in the past, based on largely superficial morphological differences and now considered invalid. Until recently, the most widely accepted taxonomy separated the Lion into Asian populations,

African (East and Southern) Lion

Asiatic Lion

P. l. persica, and African populations, *P. l. leo*. However, genetic analyses strongly suggest the primary division between lion populations actually occurs in Africa roughly either side of the equator. An estimated 178,000 to 417,000 years ago, Lion range divided as repeated cycles of extremely wet and dry periods produced massive fluctuations in the extent of the Congo Basin rainforest and Sahara Desert. Neither habitat is suitable for Lions, which are thought to have retreated to two refugia, in West and Central Africa (or perhaps Asia) and in southern Africa. These two separate populations subsequently reoccupied suitable habitat during climatically stable periods when savannas expanded, coming together only relatively recently. The resulting dichotomy, reflected by strong genetic evidence, groups Asian populations with those in Central and West Africa (which would take the subspecies name *P. l. leo*, as the type specimen was described from North Africa

within this range); and groups together populations from East and southern Africa (*P. l. melanochaita*). The same genetic analyses show evolutionarily significant subpopulations within the two main groups that are insufficiently different to be regarded as subspecies. In *P. l. leo*, there are three such units: Asia with the Middle East and North Africa (Lions are extinct in the latter two); West Africa west of the lower Niger River; and Central Africa. In *P. l. melanochaita*, there are three units: north-eastern Africa; south-western Africa; and East Africa with the rest of southern Africa.

Description

The Lion is the largest carnivore in Africa and the second largest cat species. On average, Lions are comparable to Tigers in all measurements and, in fact, the Lion has a longer skull on average (around 1.2cm longer in females and 2cm longer in males).

The foundation of Lion sociality is a matriline of related females and their cubs. The largest prides are found in mesic woodland savannas where there is a high abundance of very large prey.

The largest Tigers on record are slightly heavier and longer, while the smallest adult Tigers, from Sumatra, are considerably smaller than the smallest adult Lions. The Lion is a massive, robustly built cat with a deep chest, very powerful forequarters and heavily built limbs. The Lion is the most sexually dimorphic cat, both by secondary sex pelage (mane, belly fringe and elbow tufts in the male) and by size; male Lions are typically 30–50 per cent heavier than females. There is relatively little regional variation; Lions from India and the Sahelian savannas of West and Central Africa are 10–20 per cent smaller on average than Lions from southern and East African mesic savanna woodlands. Asiatic Lions usually have a distinct belly fold, which is only occasionally present in African Lions.

The Lion is uniformly coloured without body markings and is typically pale to dark tawny or sand coloured with cream or white underparts; lightening or deepening of the body colour produces variations including ash-grey, buff, light auburn and, rarely, dark brown. The backs of the ears are contrasting black with a scattering of silver hairs; the black ear-backs are distinct from a distance and may help Lions to locate hidden pride members spread out during a hunt. The tail ends in a distinct tassel of black or very dark brown fur, possibly useful as a flag for young cubs to follow in tall grass. White lions from the Kruger National Park region, South Africa, are leucistic (not albino) arising from a recessive gene for coat colour, and have pigmented eyes, nose and pads; they can be born to normally coloured parents. There are no records of melanistic lions. Lion cubs are born with dark brown rosettes, likely indicative of a more spotted, forest-dwelling ancestral species, which perhaps helps to camouflage young cubs in dens. The markings fade with age and are retained as faint spotting in some adults on the underparts, and, very rarely, covering the entire body (giving rise to historical accounts of new 'spotted lion' species).

The Lion is the only felid in which the males develop an extensive mane (that varies widely in

The Lion's face is almost entirely devoid of markings. Pale fur under the eyes might help to reflect light into the eye for nocturnal hunting. The pattern of whisker spots is unique to individuals although it is not thought to have any function.

colour and extent). Colour ranges from blonde to black, often with a corona of lighter colour surrounding the face. Mane growth begins at six to eight months, usually later in very hot climates. In mature males, the mane usually covers the entire head (excluding the face), neck, shoulders and upper chest; growth tends to be most extensive in mesic regions of southern and East Africa especially above 1,000m, sometimes extending along the ribcage and the belly. Captive individuals in northern hemisphere zoos with cold winters often grow extensive manes. The least developed manes occur in very hot climates, for example from Tsavo National Park, Kenya, to Niassa National Reserve, northern Mozambique (including Selous Game Reserve, Tanzania); Lions living at very low altitude here may be virtually maneless. Manelessness is also common in Sahelian populations. Male Asiatic lions have somewhat reduced manes, mainly around the face and crown. The mane is thought to signal the male's relative genetic fitness to Lionesses and to intimidate rivals – mane length and colour convey information about male aggression and the ability to defend a pride from other males. Lionesses tend to choose males with the longest and darkest manes.

Similar species The Lion is one of the most recognisable mammal species and is quite distinct from all other cats. The unicolour Puma or Mountain Lion was named for its similar colouration to a Lioness, but the resemblance is otherwise weak and their ranges do not overlap.

Distribution and habitat

The Lion has a patchy distribution in Africa south of the Sahara, chiefly in and around protected areas, and is restricted to a single Asian metapopulation in Gujarat state, India. The largest and most extensive Lion populations occur in East and southern Africa. They are severely reduced in most of Central Africa, with a fragmented distribution in northern Cameroon, southern Chad, Central African Republic,

South Sudan and northern Democratic Republic of the Congo. Lions are extinct in most of West Africa where there are now only four populations in Senegal, Nigeria and a large tri-national population on the shared borders of Benin, Burkina Faso and Niger. The Lion is now extinct in North Africa, the Middle East and Asia except for the single population of around 400 in India.

Lions occur in a broad variety of habitats. They reach highest densities in mesic, open woodland and grassland savannas but they also occupy all kinds of moist and dry savanna woodlands, dry forest, scrub savanna, coastal scrub and semi-desert including very arid environments in the Kalahari and northern Namibia; they do not inhabit the interiors of true deserts and are absent from the Sahara. They traverse forest patches in savannas, such as the montane Harenna Forest in the Bale Mountains, Ethiopia, and Maramagambo Forest in Queen

Lion sociality is thought to have arisen in open savanna habitat where a lone Lioness with a large carcass would have been vulnerable to kleptoparasitism from a wide variety of predators including many extinct species. Forming groups allowed close relatives, rather than competitors, to benefit.

Elizabeth National Park, Uganda. Until very recently, they occurred widely in mosaics of equatorial forest and savanna, for example in Gabon and Republic of the Congo, but they are naturally absent from extensive moist, dense forest including the entire Congo Basin forest. The Indian population inhabits a mosaic of dry, deciduous teak forest and Acacia savanna. They occur from sea level to 3,500–3,600m (for example, on Mount Elgon and Mount Kenya, Kenya) and are recorded exceptionally moving to 4,200–4,300m (Bale Mountains, Ethiopia, and Mount Kilimanjaro, Tanzania). Except for largely unmodified livestock areas with low human densities and extant wild prey, Lions cannot occupy human-modified landscapes.

Feeding ecology

The Lion is a highly opportunistic and formidable predator. A single adult Lion is able to overpower much larger prey and, in prides, they are able to kill virtually everything they encounter; only healthy, mature male elephants are invulnerable to Lion predation. Lions are recorded eating insects to beached whale carcasses, but populations cannot persist without large herbivores weighing 60–550kg. In any given population, the diet is typically dominated by three to five ungulate species such as zebras, wildebeests, buffaloes, Giraffes, Gemsbok, Impalas, Nyalas, Greater Kudus, kobs, Thomson's Gazelles, Chitals, Sambars and Warthogs. Small prey may dominate seasonally; for example, non-migratory Impalas and Warthogs are important prey to Lions during the lean, dry season in areas where large ungulates are migratory, including in Chobe National Park, Botswana, and Serengeti National Park, Tanzania.

Very large prey, including adult Common Hippopotamuses, rhinos and female elephants, are

Contrary to popular myth, male Lions are frequent and successful hunters. In the southern Kruger National Park, males make 60 (for territorial males) to 87 per cent (non-territorial males) of their own kills. The rest are scavenged, mainly from Lionesses.

All cats feed using their cheek teeth – the robust, sharp premolars and molars – necessitating turning the head on the side to shear through tough hide and slice off chunks of meat.

killed by large Lion prides usually when weakened through malnutrition, and a single lion is able to kill juveniles of these species. Over four years, 1993–1996, a pride in Chobe National Park, Botswana, consisting of two adult males, eight adult Lionesses and various cubs, killed 74 Savanna Elephants, including six adult females and one adult bull (previously injured by another bull). As elephant densities increase, for example in Chobe and Hwange National Parks, the proportion of elephant calves in Lion diet increases. Besides large ungulates, practically any species may be eaten by the Lion but none contributes significantly to the diet. Lions are recorded killing Aardvarks, porcupines, primates, including gorillas, Chimpanzees and baboons, many bird species including Ostriches, reptiles including large Nile Crocodiles and African Rock Pythons, fish and a wide variety of invertebrates. Lions regularly kill other carnivore species including Leopards, Cheetahs, hyaenas, African Wild Dogs, Cape Fur

Seals and a wide variety of smaller species. Carnivores are apparently unpalatable to Lions and are generally left uneaten when killed. Cannibalism occasionally occurs, most often of cubs killed by infanticidal male Lions. Although Lions go for long periods killing only wild prey even when domestic livestock is abundant in an area, they do prey on livestock including cattle, goats, sheep, donkeys, horses, camels and Water Buffaloes (India), and are occasionally recorded killing domestic dogs. Lions rarely eat people although isolated, localised pockets of persistent man-eating occur very rarely, for example in areas of south-eastern Tanzania and northern Mozambique, where Lions locally regard people as prey. An estimated 50–100 people are killed annually by Lions in remote communities in these areas.

Hunting is mainly nocturno-crepuscular and terrestrial. Lions are too heavy to climb well, and do not hunt arboreally, although prey (for example, baboons and guineafowl) is sometimes snatched from lower branches. Lions move and hunt communally, often with the entire pride in attendance, including cubs older than four to five months. Young cubs are left behind, sometimes in the company of older cubs or males. Adult females typically initiate most foraging bouts and they make the most kills. However, male Lions are capable hunters that regularly make their own kills when unaccompanied by the females and males increase the success rates of prides hunting very large prey, for example buffaloes, Giraffes and elephants. Lions search for prey as they walk, using their excellent vision and hearing to identify potential targets; or they rest around profitable ambush sites such as waterholes until a suitable opportunity arises. Most hunts begin with a careful stalk by one or more members of the pride to as close as 15m from the prey, followed by an explosive rush. The Lion's top speed is an estimated 58kmph, which it can maintain only for around 250m, and most chases are shorter. The exception occurs during hunts of buffalo herds that often stand their ground. Lions harass the herd until it stampedes, providing opportunities for the Lions to isolate a straggler during the ensuing chase, which often covers up to 3km with an exceptional record of 11km. Lions kill large prey by a suffocating throat bite administered by an adult, often while the rest of the pride begins feeding.

Lions hunt communally but the degree of actual cooperation is exaggerated. On occasion, a single Lioness stalks prey while the rest of the pride watches, joining her only when the prey is caught. More often, multiple females are involved in which they fan out around the prey and one or more individuals gives chase; the capture is often the result of fleeing prey running into a Lioness lying in wait. During hunts of very fast Springbok in open habitat in Etosha National Park, Namibia, Lionesses apparently assume a specific position and coordinate their movements, with 'wing' positions driving Springbok to 'centre' positions. Cooperative hunts are more successful than hunts by lone Lionesses, and hunts are more successful when each Lioness occupies her preferred position. Cooperative hunting is also more prevalent during hunts of large dangerous prey, especially buffaloes and elephants. Estimates of hunting success include: 15 per cent (Etosha National Park, Namibia), 23 per cent (Serengeti National Park, Tanzania) and 38.5 per cent (Kalahari, South Africa). Lions readily scavenge, and this comprises 5.5 per cent of intake in Etosha National Park to almost 40 per cent in Serengeti National Park. They frequently appropriate kills from other carnivores.

Social and spatial behaviour
The Lion is highly social and is the only felid that forms large, mix-sexed prides comprising multiple generations. Lions are almost constantly in the company of other pride members and rarely alone

The formidable African Buffalo is responsible for more injuries and fatalities to Lions than any other prey species yet it is a frequent prey item, especially for males. For male Lions in southern Kruger Park, buffaloes comprise 36 (for territorial males) to 73 per cent (for non-territorial males) of all kills. Buffaloes make up 18 per cent of Lionesses' kills.

In between feeding sessions, cubs of many cat species often exuberantly attack dead prey animals, like this five-month-old Lion cub with a Giraffe carcass. Cubs usually focus such play attacks around the head and neck, instinctively practising the throttling killing bite.

except when females temporarily abandon the pride to give birth. The basis of the pride is a matriline of 1–20 (usually three to six) related Lionesses that communally defend a territory and raise their cubs. Each pride usually has a coalition of one to nine (usually two to four) adult males that typically immigrate from other prides and are unrelated to the breeding females. Pride size is only weakly correlated to prey abundance (whereas this strongly influences home range size and Lion density) and is remarkably consistent over much of the East and southern African range – typically three to six females and two to three males. This would probably also apply in West and Central Africa if it wasn't for very high levels of human persecution of Lions and prey. Pride size exceptionally and temporarily reaches 45–50 including cubs in optimal conditions but invariably decreases as large cohorts of subadults disperse.

Female membership of the pride is stable but small subgroups often come and go within the pride range in a 'fission-fusion' pattern so that the entire pride is rarely all together. Females usually stay with

the pride for life. They occasionally disperse following a pride takeover by immigrant males, or to avoid mating with male relatives. Young males are evicted or leave the pride at 20–48 months, entering a protracted transient phase lasting up to three years before attempting to acquire their own pride. Coalition males are usually related to each other but male singletons and pairs often join with unrelated males during dispersal to form a coalition. Whether related or not, coalition members have close bonds and remain together for life, defending their territory and females from male intruders. Immigrant males challenge resident males for access to female prides, sometimes resulting in violent fights that are often fatal. Lionesses participate in pride defence and large female groups may succeed in repelling incoming coalitions, though Lionesses are occasionally killed by strange males in these encounters. Inter-pride territorial skirmishes are also common, sometimes resulting in neighbouring Lionesses killing each other or cubs. When new males succeed in taking over a pride, they usually kill or evict all unrelated cubs younger than 12–18 months to hasten the Lionesses' return to oestrus. Depending on coalition strength, males may take over a neighbouring pride while still defending their current pride. Coalition tenure is generally two to four years.

Territories are very stable, with female matrilines remaining in essentially the same area for many generations. Territory size varies widely, depending on the productivity of the habitat and therefore the available prey biomass. Pride territories in Gir Protected Area (India) where there is a superabundance of Chital deer are very small at 12–60km². Serengeti (Tanzania) pride territories average 65km² (woodlands) to 184km² (grasslands), reaching a maximum of 500km². Average range size in West and Central Africa (based on very few prides) ranges from 256 km² in relatively well-watered Pendjari National Park, Benin, to 756km² in drier Waza National Park, Cameroon. In semi-arid savanna in Hwange National Park, Zimbabwe, female prides have an average range size of 388km^2 (range, 35–981km^2) and slightly more for male coalitions, averaging 478km^2 (71–1002km^2). Range size is very large in arid areas; for example, 1,055–1,745km^2 (Khaudom Game Reserve, Namibia), 266–4,532km^2 (Kgalagadi Transfrontier Park, South Africa) and 2,721–6,542km^2 (Kunene, north-western Namibia). Two Kunene male coalitions (possibly nomads) had ranges of 13,365–17,221km^2. Density estimates include 0.05–0.62 per 100km^2 (Kunene, Namibia), 1.5–2.0 per 100km^2 (Kalahari, South Africa), 3.5 per 100km^2 (Hwange National Park, Zimbabwe), 6–12 per 100km^2 (Kruger National Park, South Africa), 12–14 per 100km^2 (Gir Protected Area, India) and up to 38 per 100km^2 (Lake Manyara National Park, Tanzania).

Reproduction and demography

The Lion breeds aseasonally, though births often peak when seasonally breeding ungulates give birth, for example March–July in East Africa (Serengeti National Park, Tanzania) and February–April (Kruger National Park, South Africa). Oestrus averages four to five days and gestation is 98–115 days, averaging 110 days. Litter size is typically two to four cubs, exceptionally up to seven. Lionesses usually leave the pride to give birth, and cubs are kept isolated until around six to eight weeks old, perhaps to insulate them from rough play or accidental deaths by other pride members. Females in the same pride often give birth synchronously and communally care for cubs; females suckle all cubs but they only transport their own by holding them gently in the mouth. Weaning begins around six to eight weeks

An Asiatic Lioness and cubs in the Gir Forest, India. Persecution and trophy hunting had reduced this population to around 25 individuals by the start of the 20th century. Strict protection imposed in the early 1900s has resulted in a spectacular population rebound where now the challenge is a lack of habitat for further expansion.

but suckling may continue to eight months. Inter-litter interval is typically about three years in savanna-woodland ecosystems with resident prey, and declines to 20–24 months in areas with migratory prey, where cub survival rates are consequently lower. When reintroduced into areas from which they have been previously extirpated, Lions breed at similarly high tempos to migratory systems. Cubs can hunt independently at around 18 months but they rarely disperse before two years, and typically leave the pride at just under 36 months of age when their mothers initiate the next breeding event (in most savanna-woodland habitats).

In migratory prey systems and arid habitat, dispersers wander extensively, often ending up more than 200km from their natal range. However, in wooded savannas with abundant resident prey (especially buffalo), most young males remain within or close to their natal range for several years and may establish territories close to their original pride. Males are more likely to survive dispersal if they leave the pride as old as possible. All dispersers younger than 31 months monitored between 1999 and 2012 in Hwange National Park, Zimbabwe, did not survive. Of 49 dispersers whose fates were known, 65 per cent (17 males and 15 females) survived to establish a territory, while 26 per cent were killed by people. A male Lion in this population that dispersed at 27 months was transient for 848 days and travelled 4,223km before he was shot for killing livestock. Lionesses can conceive at 30–36 months but typically first give birth around 42–48 months and cease reproducing after age 15. Males are sexually mature at 26–28 months but rarely breed before five to six years of age. In areas where adult males are depleted (typically due to excessive

Male Lions are tolerant and affectionate fathers to their own cubs and their presence is crucial to cub survival. Males constantly patrol their territory against incursions from intruding males which kill unrelated cubs.

trophy hunting), very young males of two to four years replace these older mature males as the main breeding males in the population.

Mortality Cub mortality varies widely, depending on the region and seasonal or annual fluctuations in prey availability. In the first year, around 16 per cent of cubs die in southern Kruger National Park, South Africa, where there is abundant, resident prey. In the arid Kalahari, this figure is around 40 per cent. It is highest in areas with dramatically fluctuating prey levels; for example, with migratory systems such as in Serengeti National Park, Tanzania, where an average of 63 per cent of cubs die in the first year. Cubs die mostly from infanticide (which is higher in the Serengeti than anywhere else), predation (mainly by Leopards and Spotted Hyaenas) and starvation due to food shortages. Annual cub mortality in the second year drops dramatically; for example, to 20 per cent (Serengeti) and 10 per cent (Kruger). Male Lions experience much higher rates of mortality during dispersal, resulting in a typical adult sex ratio of two to three Lionesses per adult male. Apart from human-caused mortality, adult Lions die mainly in fights with other Lions (especially males), from injuries sustained while hunting large prey (especially buffaloes) and from starvation when old or debilitated. Lions often survive catastrophic injury from fights or hunting accidents, usually to the lower spine and hindlegs. These Lions seldom live to an old age. Disease is uncommon but the intense sociality of Lions fosters disease transmission, and episodes are occasionally severe; more than 1,000 lions (40 per cent of the population) died during a 1993–4 canine distemper outbreak in the Serengeti.

Lifespan 18 years for females, 16 (but rarely over 12) for males in the wild and 27 in captivity.

STATUS AND THREATS

The Lion reaches high densities in well-protected, productive habitat, but it has undergone a massive range collapse and continues to decline across most of its remaining distribution. The species is extinct in Asia except in Gujarat, western India, where about 300 live in the Gir Protected Area (1,883km²) and another approximately 100 live in seven small satellite populations nearby. In Africa, the most optimistic estimate of Lion distribution is 16.3 per cent of its historical range, counting poorly known areas where its continued presence is uncertain but possible. The Lion is unequivocally known from slightly less than 8 per cent of its original range. It is extinct in 15 African countries and possibly extinct in an additional seven. The latest estimate of approximately 32,000 Lions is overly optimistic; the number is probably closer to 20,000, many of which occur in small, isolated and declining populations. There are only eight countries thought to contain at least 500 adults: Botswana, Kenya, Mozambique, Namibia (possibly), South Africa, Tanzania, Zambia and Zimbabwe. The Lion is Critically Endangered in West Africa where the number of adults is thought to be fewer than 250.

The Lion continues to decline over most of its range. Overall, the total population is estimated to have declined 38 per cent since 1993 but this conceals a far more severe decline across most of the range. Five countries (Botswana, India, Namibia, South Africa and Zimbabwe) have stable (or nearly stable) and increasing populations where the total population is estimated to have increased 25 per cent since 1993. Such increases in a relatively small part of the range disguise the severity of the decline everywhere else, which is estimated at 59 per cent since 1993. Accordingly, although the Lion is classified as Vulnerable, it qualifies to be considered Endangered in most of its range.

Lions have declined mainly from the extensive loss of habitat and prey to agriculture and livestock herding, coupled with pervasive eradication by people. They are persecuted intensely by herders and suffer from widespread official killing of 'problem' animals. Their habit of scavenging makes them highly vulnerable to poisoned baits and to snares set for bushmeat. Sport-hunting is legal in 13 African countries, taking approximately 600–700 (mainly males) annually, and contributes to population declines when poorly managed. Genetic impoverishment of small, isolated populations possibly leads to declines and vulnerability to disease.

CITES Appendix II. Red List: Vulnerable (global), Endangered (Asia), Critically Endangered (West Africa). Population trend: Decreasing.

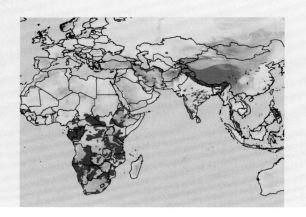

IUCN RED LIST (2008):
- Near Threatened (global)
- Endangered (Sri Lankan Leopard, Persian Leopard)
- Critically Endangered (Javan Leopard, Arabian Leopard, Amur Leopard)

Head-body length ♀ 95–123cm, ♂ 91–191cm
Tail 51–101cm
Shoulder height 55–82cm
Weight ♀ 17.0–42.0kg, ♂ 20.0–90.0kg

Leopard

Panthera pardus (Linnaeus, 1758)

Panther

17–28.2cm

Melanistic form

African Leopard
(forest form)

Arabian Leopard

Amur Leopard

African Leopard
(savanna form)

Taxonomy and phylogeny

The Leopard is one of the 'big cats' in the genus *Panthera* and is most closely related to the Lion and Jaguar. Molecular analyses suggest one subspecies in Africa (African Leopard *P. p. pardus*) and eight Asian subspecies, though an up-to-date, comprehensive analysis is long overdue – Arabian Leopard *P. p. nimr* (Middle East), Persian Leopard *P. p. saxicolor* (central Asia), Indian Leopard *P. p. fusca* (Indian sub continent), Sri Lankan Leopard *P. p. kotiya* (Sri Lanka), Indochinese Leopard *P. p. delacouri* (South-east Asia to southern China), Javan Leopard *P. p. melas* (Java), North Chinese Leopard *P. p. japonensis* (northern China) and Amur Leopard *P. p. orientalis* (north-eastern China and Russian Far East). There is recent evidence that *fusca*, *kotiya* and western populations of *delacouri* represent the same subspecies, and similarly for eastern *delacouri*, *japonensis* and *orientalis*, which would reduce the number of Asian subspecies to five.

Left: **Relative to smaller felids, the front of the skull in large cats like the Leopard is enlarged, with elongated, massive jaws and their associated musculature to deal with large prey.**

Below: **Melanistic Leopards are spotted just like their normal-coloured counterparts but the spots are only visible under oblique light. Biologists have discovered that infrared camera-trap photography reveals the spots, allowing black individuals to be identified by their unique coat pattern,**

Description

The Leopard is a large, robust cat with very muscular forequarters, slender hindquarters and a long tail around two-thirds of head–body length. The head and neck are heavily built, especially in adult males, which also often develop a distinctive throat dewlap by the time they reach five years old. The Leopard's body size varies widely and correlates with changes in climate and prey availability. The smallest Leopards, from arid mountainous areas in the Middle East, are about half the weight of African savanna-woodland individuals. Leopards from an isolated population in the coastal Cape Fold Mountains, South Africa, are also very small, averaging 21kg (females) to 31kg (males). The largest Leopards are recorded from East and southern African woodlands, and central Asian temperate forest.

The Leopard's background colour is highly variable, including various shades of creamy yellow, buff-grey, ochre, orange, tawny-brown and dark rufous-brown. The underparts are creamy to bright white. The body is covered with densely packed rosettes, each a cluster of small, black spots around a normally unspotted centre that is slightly darker than the body colour. The spots become large solid blotches on the lower limbs, belly, tail and throat, often merging into a distinctive yoke on the chest. The overall colour is broadly associated with region and climate, paler in arid and temperate areas, and darker in dense habitat and more tropical areas. Melanism is an inherited recessive trait which occurs widely, usually associated with humid subtropical, tropical and montane forests, and occasionally from drier woodland habitats (e.g. Laikipia Plateau, central Kenya). Melanistic individuals ('black panthers') are most common in tropical southern Asian populations, especially in Malaysia and Java. Surveys of Ujung Kulon National Park, Java, in 2000–2003 gathered 40 melanistic and 69 spotted Leopard photographs. Melanistic individuals accounted for 445 of 474 Leopard camera-trap photographs from across Peninsular Malaysia and southern Thailand. All Leopards photographed south of the Kra Isthmus were melanistic, although spotted individuals occasionally occur. In Africa, melanism is mostly associated with montane forest refugia (e.g. Aberdare Mountains and Mount Kenya, Kenya; Virunga Mountains, East Africa; and Harenna Forest, Ethiopia) and not the Congo Basin rainforest as widely assumed. Erythristic and albino individuals are occasionally recorded, as is wide variation in spotting ranging from overall diminution to give a peppered or freckled appearance, to extensive marbling and coalescing of the spots (pseudomelanism or abundism) similar to the King Cheetah (p169).

Similar species The Leopard closely resembles the Jaguar but they are not sympatric in the wild; apart from a larger, much heavier overall appearance, Jaguars have large, blocky rosettes with internal spots rarely found in Leopards. The Cheetah is roughly similar in overall size and colouration to the Leopard but the two should not be confused.

Distribution and habitat

The Leopard has the largest distribution of all wild cats. It is widely but patchily distributed in southern, East and Central Africa, rare in West Africa and relict or extinct in North Africa. It is largely relict or extinct in Turkey, Georgia, Azerbaijan and the Arabian Peninsula; the most viable populations inhabit the Dhofar Mountains, Oman, and Wada'a Mountains, Yemen. In central Asia, it is rare and now widely distributed only in Iran at low densities. They occur widely but patchily throughout South and South-east Asia including Sri Lanka and Java (extremely rare) into south-eastern China with an isolated population in East China and the Russian Far East. Leopards naturally never occurred on Bali, Borneo or Sumatra.

Leopards have a very broad habitat tolerance, ranging from Russian deciduous forest with winter lows of -30°C, to deserts with summer highs exceeding 50°C. They reach highest densities in mesic woodland, grassland savanna and subtropical to tropical dry and humid forest. They are fairly

common in mountains, temperate forest, shrubland, scrub and semi-desert. They are absent from the open interiors of true desert but they inhabit watercourses and rocky massifs in very arid areas. They tolerate human-modified landscapes provided cover and prey are available, including coffee plantations, fruit orchards and irrigated agricultural landscapes, sometimes densely populated with people such as cropland-dominated valleys in Maharashtra, India. Leopards occur from sea level to 4,200m and are recorded exceptionally up to 5,200m (Himalayas). A dead Leopard was found at 5,638m on Mount Kilimanjaro, Kenya.

Feeding ecology

The Leopard is renowned for its ability to exploit an extremely wide variety of prey species. In concert with its vast range and very broad habitat tolerance, this results in the Leopard having the most diverse diet of any felid. Counting only mammals weighing more than 1kg, at least 110 species are recorded as prey, and including all vertebrates, i.e. small mammals, birds, herptiles and fish, the total exceeds 200 species. Notwithstanding its very catholic diet,

ungulates weighing 15–80kg form the mainstay of Leopard diet, with one to two locally abundant herbivore species making up most of the diet in any given Leopard population; for example, Impala (48–93 per cent of kills) in southern African mesic savannas, Springbok (65 per cent of kills) in dry Kalahari savanna, Nyala (43 per cent of kills) in KwaZulu-Natal dense woodland savanna, Axis Deer (more than 50 per cent of kills) in Sri Lankan dry forest, and Siberian Roe Deer and Sika Deer (more than 50 per cent of kills) in Russian deciduous forest. Other common prey includes Steenboks, duikers, muntjacs, gazelles, Bushpigs and Warthogs; and juveniles of larger or dangerous species such as wildebeest, oryx, Hartebeest, Greater Kudu, Giraffe, Sambar, Gaur, Asiatic Water Buffalo and Wild Boar. Exceptional prey includes very young elephants and rhinos, as well as subadults and adults of very large ungulates; the largest on record is an adult male Eland (approximately 900kg), a testament to the formidable strength of the Leopard. The Leopard's reputation for preferring primates is exaggerated; primates are most important to African rainforest Leopards where at least 15 species are recorded as

These large cubs at play in Ruhuna National Park, Sri Lanka are close to the age at which they will disperse. Like most felids, they will be solitary as adults but not asocial; familiar females and males interact frequently in a complex social system of friendly relationships and rivals to be avoided.

By focusing on large-bodied prey, often their own size or larger and equipped to defend themselves with horns or tusks, large felids risk serious injury. This Sri Lankan Wild Boar repelled the Leopard's attack, fortunately without injury to the cat.

prey, but these are still secondary to ungulates. Guenons (genus *Cercopithecus*), colobus monkeys (genus *Colobus*), vervets (genus *Chlorocebus*) and langurs (genus *Semnopithecus*) are typical primate prey but Leopards occasionally kill various baboon species, Chimpanzees, Bonobos and Lowland Gorillas. Leopards will opportunistically kill anything they can catch or overpower; hyraxes, hares, rodents and birds are often locally important. They are recorded killing very large (up to 4m) African Rock Pythons and Nile Crocodiles up to 2m. Other carnivores are readily killed and sometimes eaten, including adult Spotted Hyaenas, Cheetahs and Lion cubs (rarely); e.g. an adult male killed and hoisted two cubs from the same Lion pride in separate incidents (and was later killed by the same pride; Sabi Sands Game Reserve, South Africa). Cannibalism occurs rarely, typically of cubs killed by males and sometimes of adults killed during intraspecific fights. Leopards prey on livestock, occasionally entering corrals and settlements, and they readily kill domestic dogs. Dogs followed by

domestic cats and cattle were the most important prey items (by biomass) for a Leopard population occupying highly modified agricultural habitat in Maharashtra, India; 87 per cent of the diet was made up of domestic animals. Leopards sometimes prey upon people, though habitual man-eaters are very rare.

Leopards forage alone, including mothers with large cubs that are left behind. Most hunts are nocturno-crepuscular; daylight hunts are likely to be opportunistic and are generally less successful. The Leopard is a consummate stalk-ambush hunter that varies hunting strategy according to the prey species and habitat. In open habitat such as in northern Namibia and the Kalahari, most hunts are preceded by careful stalks averaging 29–196m until the Leopard is less than 10m from the target. Long stalks are rare in dense habitats with limited visibility, and sit-and-wait ambush hunting appears to be more common. African rainforest Leopards hide in dense vegetation near monkey groups, presumably waiting for them to move close, or they position themselves where encounters with potential prey are likely; for example, in trailside vegetation along game trails or at fruiting trees that attract primates, duikers and Red River Hogs. Most prey species, especially larger ungulates, are killed by asphyxiation with a throat bite. A suffocating muzzle bite is sometimes used, for example with oversized prey; a 14-month-old female Leopard at Phinda weighing 20kg took 14 minutes to suffocate a mature male Impala (60kg) by this method. Leopards normally quickly dispatch smaller prey by biting the back of the skull or neck.

Estimates of hunting success include 15.6 per cent (open Kalahari savanna), 20.1 per cent (dense woodland savanna, Phinda, South Africa) and 38.1 per cent (open, arid savanna, northern Namibia). Daylight hunts in the Serengeti National Park (Tanzania) are estimated to have a 5–10 per cent success rate. In the southern Kalahari, mothers with cubs have higher success (27.9 per cent) compared with females without cubs (14.5 per cent) and males

(13.6 per cent). Kalahari mothers kill more smaller prey than lone females or males but travel less for each kill and they kill more often. Kalahari Leopards make an average of 111 (males) to 243 (females) kills annually. Leopards use their tremendous strength to hoist carcasses into trees to avoid kleptoparasitism, including observations of a young Giraffe (estimated at 91kg) and a month-old Black Rhino. The behaviour is apparently most common in African savanna woodlands, but records exist from throughout the range including separate incidents where kills were hauled in response to the presence of Dholes and a Striped Hyaena in India. Leopards occasionally also cache in caves, burrows and kopjes. Leopards in open habitat seek out dense brush or scrub in which to hide kills, and similarly use dense vegetation on the ground in areas where competition from Lions and Spotted Hyaenas is low. They typically pluck fur or feathers (from large birds) prior to feeding, usually starting at the underbelly or hindlegs. Leopards

readily scavenge and appropriate kills from other species including Cheetahs, African Wild Dogs, jackals and hyaenas.

Social and spatial patterns

Leopards are solitary; adults socialise mainly when mating but males frequently associate amicably with familiar females and cubs, including sharing carcasses. Males are tolerant of cubs belonging to females they have mated. Both sexes maintain enduring territories in which the ranges of males are large and typically overlap one or more, smaller ranges of females. Adults defend a core area against same-sex conspecifics but they tolerate considerable overlap at range edges, using a 'timeshare' pattern of mutual avoidance and alternating use of shared areas to avoid conflict. Territorial fights are uncommon, especially between long-term neighbours that are aware of each other. Serious fights are more likely to occur when an immigrant adult moves into an area,

Despite a reputation for hunting in the densest available habitats, Leopards in woodland savannas actually prefer areas with intermediate cover, even when prey is more abundant in dense habitat. Intermediate habitat probably provides the best balance between finding prey and being able to catch it.

The Leopard is the only felid to have evolved the strategy of routinely hauling carcasses into trees to avoid losing them to other carnivores. It might have arisen during the Pleistocene when Leopards shared the landscape with extinct sabre-toothed cats and giant hyaenas in addition to modern competitors like Spotted Hyaenas and Lions.

or when the balance of power shifts between residents, for example if one is injured. Such fights sometimes result in fatalities in both females and males.

Both sexes demarcate territory and advertise reproductive availability by depositing scent-marks via cheek-rubbing and spraying vegetation, and by depositing scats and scraping the ground with the hindfeet. The Leopard's most distinctive call, sawing (also called coughing or rasping), carries up to 3km and probably serves the same dual purpose. Leopards can almost certainly recognise familiar individuals by this call although this is yet to be tested. Scent-marks are deposited liberally along frequently used routes including game paths, trails, roads and along territory boundaries. A territorial male Leopard continually followed for 65 minutes on 'patrol' in Phinda sprayed 17 times, scraped 6 times and sawed 5 times.

Territory size varies with habitat quality and prey availability, ranging from 5.6km² (female, Tsavo National Park, Kenya) to 2,750.1km² (male, southern Kalahari). Mean range size for mesic woodlands, savannas and rainforest averages 9–27km² (females) and 52–136km² (males). Ranges are much larger in arid habitats, averaging 188.4km² (females) and 451.2km² (males) in northern Namibia, and 488.7km² (females) and 2,321.5km² (males) in the Kalahari. A collared male in arid, rocky habitat, central Iran, used 626km² in 10 months. Leopards reach high densities in well-protected, high-quality habitat. Counting only estimates from protected areas, density varies from 0.5 leopards per 100km² (Etosha National Park, Namibia) to 1.3 per 100km² (Kalahari), 3.4 per 100km² (Manas National Park, India), 11.1 per 100km² (Phinda–Mkhuze game reserves, South Africa), 13 per 100km² (tr opical deciduous forest, Mudumalai Tiger Reserve, India) and 16.4 per 100km² (southern Kruger National Park). In African woodland-savannas, average density under protection (10.5 per 100km²) is almost five times as high as outside protected areas (2.1 per 100km²). Density under strict protection in Gabon rainforest is 12 per 100km² compared to 4.6 per 100km² in heavily exploited forest. Amur Leopards occur at approximately 1–1.4 per 100km² in deciduous forest that is mostly poorly protected (Primorsky Krai, Russia), and Leopards occupying heavily converted and densely populated habitat outside protection in western Maharashtra, India, reach 4.8 per 100km².

Reproduction and demography

The Leopard breeds year-round but may exhibit a strong birth pulse negatively associated with the lean season; e.g. twice as many litters are born in north-east South Africa in the wet season (October–March, when seasonally-breeding ungulates give birth) compared to the dry season. This is poorly known in their northern range where extreme winters are likely to give rise to seasonality, e.g. in northern Iran and the Russian Far East. Oestrus lasts 7–14 days, during which females call and scent-mark constantly, and sometimes make long forays outside their territory to find males, e.g. up to 4.7km in Phinda and Mkhuze game reserves (South Africa).

Gestation is 90–106 days. One to three cubs are born; litters numbering up to six are recorded in captivity but this is exceedingly rare in the wild, e.g of 253 litters born in Sabi Sands Game Reserve, none exceeded three cubs. Weaning begins around 8–10 weeks and suckling typically ceases before four months. Adoption occurs rarely, and usually between related females, e.g. a 15-year-old female adopted the seven-month-old male cub of her nine-year-old daughter and raised him to independence (Sabi Sands Game Reserve). Inter-litter interval averages 16–25 months. Cubs reach independence typically at 12–18 months; the earliest at which cubs survive is seven to nine months. Female dispersers often inherit part of their mother's range while males disperse more widely, e.g. 113km (Kalahari) to 162km (north-east Namibia). A male dispersing from Phinda Game Reserve, South Africa, entered southern Mozambique and covered a minimum distance of 356km in 3.5 months, before dying in a wire snare in north-east Swaziland, a straight-line distance of 195km from his natal range. Both sexes are sexually mature at 24–28 months; females first give birth at 33–62 months (Sabi Sands Game Reserve) at an average age of 43 (Phinda) to 46 months (Sabi Sands), and can reproduce to 16 years (19 in captivity). Males first breed around 42–48 months.

Mortality An estimated 50 per cent (Kruger National Park) to 62 per cent (Sabi Sands Game Reserve) of cubs die in the first year. Survival of cubs to independence (18 months) is 37 per cent (Sabi Sands Game Reserve). The leading cause of death in well-studied populations is infanticide by male Leopards which may kill unrelated cubs up to 15 months old; in a well-protected, stable, high-density population in South Africa, infanticide resulted in the deaths of 29 per cent of 280 cubs born between 2000 and 2012 (Sabi Sands Game Reserve). Females

Female Leopards will spend most of their adult lives raising their cubs. One intensively monitored South African female who was observed raising 10 successive litters over a 12-year period was without dependent cubs for only 22 per cent of the time.

aggressively defend cubs from infanticidal males and sometimes succeed in saving cubs but occasionally this can result in the death of the mother. Second to infanticide, most cub deaths are due to interspecific predation, mainly by Lions and less so Spotted Hyaenas in Africa, but not much is known about Asia. Estimates of adult mortality include 18.5 per cent (Kruger National Park) to 25.2 per cent (Phinda). Aside from deaths by people, adults are killed primarily in territorial fights with other Leopards and by other carnivores, especially Lions and Tigers; over 21 months, Tigers killed three adult leopards and two cubs in 7km^2 of Chitwan National Park (Nepal) in an area of high Tiger density. Groups of Spotted Hyaenas, African Wild Dogs, Dholes (and perhaps Grey Wolves although there are no observations), Wild Boar and baboons occasionally kill Leopards, usually young or infirm animals; an adult male Leopard in Ruhuna National Park (Sri Lanka) injured by another Leopard was later killed by three Wild Boar adults. Large Nile Crocodiles and African Rock Pythons are recorded rarely preying on Leopards, and there is at least one record each of cubs killed by Chimpanzees, Honey Badger, Martial Eagle and Mozambique Spitting Cobra. Hunting accidents are rare but Leopards occasionally incur fatal injuries from dangerous prey; for example, an adult male died after being gored by a Warthog. Leopards have died from wounds inflicted by crested porcupines in both Africa and Asia, though Leopards frequently kill them and typically recover from embedded spines. Deaths from disease are uncommon in Leopards. **Lifespan** Up to 19 (females) and 14 (males) years in the wild, 23 in captivity.

The skin of a poached Leopard, confiscated by wildlife authorities in Odzala National Park, Republic of Congo. Despite international trade in spotted cat fur being outlawed decades ago, cat skins are still sought after in some communities and illegal trade is widespread.

STATUS AND THREATS

For a large cat, Leopards are surprisingly tolerant of human activity. They are able to persist close to people and in human-modified habitats where other large carnivores such as Lions, Tigers, Grey Wolves and Spotted Hyaenas have long been extirpated. Even so, the Leopard has been wiped out from large parts of its historical distribution, at least 40 per cent of its African range and more than 50 per cent of its Asian range. The Leopard is common and secure in large parts of southern, East and Central Africa, and widespread but patchily distributed and less secure in much of South and South-east Asia. It is patchily distributed and rare in West Africa; Endangered in Sri Lanka (Sri Lankan Leopard: fewer than 900 remaining) and central Asia (Persian Leopard: 800–1,000 remaining); and Critically Endangered in Russia (Amur Leopard: around 60 remaining), Java (Javan Leopard: fewer than 250 adults remaining) and the Middle East (Arabian Leopard: fewer than 200 remaining).

Loss of habitat and prey is the chief threat, especially compounded by intense persecution in livestock areas and killing by people in human-dominated landscapes; at least one Leopard a week is killed by rural and peri-urban people in India, usually pre-emptively out of fear and occasionally in direct retaliation for killing livestock (or, rarely, humans). The species is heavily hunted in South Asia for skins and parts supplying the Asian medicinal trade, and is killed for skins, canines and claws in West and Central Africa. Bushmeat hunting, especially in tropical forest, competes directly for the Leopard's principal prey species and may drive extinctions even in intact forest. There is illegal but open trade of Leopard skins by members of the 'Shembe' Nazareth Baptist Church in KwaZulu-Natal, South Africa, for *amambatha* (shoulder capes). At least 1,000 *amambatha*, representing a minimum of 500 Leopards, are worn at large Shembe religious gatherings. CITES Appendix I permitting 12 African countries to export sport-hunting trophies (2013 total quota: 2,648), in addition to small numbers of live animals and skins sold commercially mainly as tourist souvenirs. Red List: Near Threatened (global), Endangered (Sri Lankan Leopard, Persian Leopard), Critically Endangered (Javan Leopard, Arabian Leopard, Amur Leopard). Population trend: Decreasing.

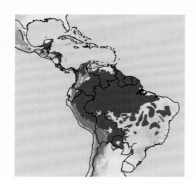

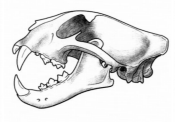

20.4–30.6cm

IUCN RED LIST (2008): Near Threatened

Head-body length: ♀ 116–219cm,
♂ 110.5–270cm
Tail: 44–80cm
Weight: ♀ 36.0–100.0kg, ♂ 36.0–158.0kg

Jaguar

Panthera onca (Linnaeus, 1758)

Taxonomy and phylogeny

The Jaguar is one of the 'big cats' in the genus
Panthera and is thought to be most closely related to
the Leopard and the Lion with a common ancestor
estimated at approximately 3–3.5 million years ago.
Historically, eight Jaguar subspecies were
recognised, based chiefly on superficial variation
in the skull, but later analysis concluded this was
insufficient to distinguish populations. Similarly,
genetic analysis has shown relatively little
differentiation among Jaguars and instead suggests
there are gradual genetic changes across a cline from
north to south without strong boundaries between
populations. Not surprisingly, the greatest
differences are found between Jaguar populations
at the latitudinal extremes of their range. The
same genetic analysis shows four weakly separated
regional groups that display some genetic
partitioning but not to the extent normally
considered to distinguish subspecies. The main
division is along the Amazon River, with the
largest group comprising all Jaguar populations
south of the Amazon. Populations north of the
Amazon are divided into three weakly defined
groups, northern South America, southern Central
America and Guatemala-Mexico. It is thought that
the Jaguar underwent a recent and rapid population

Central American form

Melanistic form

Pantanal form

expansion across the Americas approximately 300,000 years ago, after which there were few barriers impeding constant exchange of genetic material across the range. Even the Amazon and the Andes have apparently been insufficient to completely isolate Jaguar populations. Importantly, both morphological and genetic analyses are based on very small samples across a massive, continental range, and somewhat greater differentiation between populations may be uncovered with more detailed analysis.

Description

The Jaguar is the world's third largest cat and is the most robustly built of all living cats, analogous only to extinct sabre-toothed felids such as *Smilodon*. The thickset body has very muscular forequarters, a very deep chest and foreshortened waist. The limbs are short and stout, such that shoulder height is comparable to or only slightly greater than the much more lightly built Leopard. The feet are very broad and rounded with distinctly stubby and splayed digits, especially in the forefeet that act effectively to spread weight on sodden ground and as swimming paddles. The tail is relatively short compared to other large cats, around half the head-body length. The head is short, rounded and massively built, such that males especially have an unusually heavy-headed 'Pit Bull' appearance. The Jaguar's body size is broadly correlated with changes in latitude across a north–south gradient. The smallest Jaguars occur in Mesoamerican forest, with a weight range of 36–51kg

The Jaguar is the most aquatic of all the big cats and inhabits some areas which are seasonally inundated for months on end such as the Brazilian Pantanal (shown here) and the flooded forest habitat known as várzea in the Amazon.

(females) and 48–66kg (males) for animals in Mexico and Belize. The largest Jaguars inhabit the wet, savanna woodlands of Brazil (Pantanal) and Venezuela (Los Llanos) with a weight range of 51–100kg (females) and 68–158kg (males). Animals in similar habitat in this region (for example, from the Bolivian and Paraguayan Pantanal) are likely to be similarly large but measurements are mostly lacking.

The Jaguar's background colour varies in shades of buff-grey, yellow, cinnamon and tawny-orange with crisp white or creamy-white underparts. The body is covered with large, black block-like markings or rosettes with a darker brown interior and usually small black spots inside; small black spots are sometimes also interspersed between the large markings. The lower limbs and underparts are covered in large, solid black spots and blotches while smaller solid spots cover the shoulders, head and face. The ears are short and rounded with black backs and an off-white central patch. Melanism occurs as a dominant inherited trait, with the same pattern of rosettes apparent in oblique light. Melanistic individuals are most common in lowland tropical forest and are less common north of the Amazon River. The incidence of melanism appears to decline towards the periphery of the range, such that it is not known from the Atlantic Forest biome in northern Argentina and south-western Brazil.

Similar species The Leopard has a close resemblance, though the two species are not sympatric in the wild. The Leopard is less heavily built with a less massive, blocky head, and has smaller, rosette-type markings typically lacking the distinctive interior spots found in the Jaguar. The unicolour Puma is the only sympatric large cat and is never melanistic; the only large black Latin American cat is the Jaguar.

Distribution and habitat

The Jaguar occurs from northern Mexico to northern Argentina. It is widely and mostly continuously distributed in northern and central South America

east of the Andes, from the northern Andes in Colombia to the Brazilian highlands; it is patchily distributed at the eastern and southern extent of the range in northern Argentina, south-eastern Brazil and Paraguay. Its range in Central America is fragmented with tenuous links between large forested tracts that largely follow the mountainous cordilleras and associated lowlands along the Caribbean coast. The range is largely contiguous in southern Mexico and bifurcates going north, following the Sierra Madre Oriental and Sierra Madre Occidental mountain ranges. Resident, breeding populations no longer occur in the US, but individuals intermittently appear in border areas of Arizona and New Mexico from the northernmost breeding population 150km south of the border in Sonora, Mexico; three individuals, all of them male, have been documented in the US since 2001. The last record of a wild female Jaguar in the US is from 1963. The Jaguar is extinct in El Salvador and Uruguay.

The Jaguar occurs in a variety of forested and wooded habitats. It reaches highest densities in dense subtropical and tropical lowland forest, and in seasonally flooded savanna woodlands (for example,

A rare photograph of a wild melanistic Jaguar, taken in the Amazon forest, Yasuni National Park, Ecuador. Melanism occurs in at least 14 species of wild cats, the adaptive significance of which is still unknown. The trait may be adaptively neutral in which it carries no significant advantage or disadvantage.

A female Jaguar with her five-month-old cub. Compared to male Jaguars, females seem less tolerant of human-modified habitats. Radio-collared females in Mexico's Calakmul Biosphere Reserve prefer habitat with fewer roads and that is less impacted by cattle ranching and agriculture than habitat used by males.

the Pantanal), but it also occurs widely in dry forest, moist and arid areas, dense scrubland, wooded grasslands and mangrove swamps. Jaguars are closely associated with water and inhabit sodden and seasonally inundated habitats; they are excellent swimmers that routinely traverse rivers including greater than 2km widths of the Amazon and Japurá Rivers. They easily cross the Panama Canal. In arid parts of the range (for example, northern Mexico–southern US and eastern Brazil), Jaguars occur in vegetated scrubland and dry woodland associated with watercourses and mountainous terrain. Jaguars shun open landscapes with poor cover including most grassland habitats, though they occur in forested patches and riverine areas throughout some grasslands. Jaguars occur from sea level typically to 2,500m, and are rarely found in montane forest at higher elevations. They are not found above 2,700m in the Andes nor do they occur on central Mexico's high plateau. With the exception of extensive ranching areas with habitat and wild prey (such as the Pantanal), Jaguars do not occupy heavily modified, anthropogenic habitats. They have been recorded in pine and oil-palm plantations that retain

forested fragments or are close to forested habitat.

Feeding ecology

Jaguars have a diverse diet with at least 86 recorded prey species. Like all large cats, they focus on common, large-bodied prey and they are able to kill the largest available mammals, both native (Marsh Deer and tapirs) and introduced (cattle). However, the natural absence of high densities of large deer, antelope or wild cattle species in Latin America means that Jaguar diet typically includes greater proportions of both smaller species and of non-mammals than in other large cats. Across the range, Capybaras, Collared Peccaries and White-lipped Peccaries are usually the most important wild prey species where they occur in sufficient numbers; the distribution of these three species collectively overlaps that of the Jaguar almost exactly. The northern periphery of Jaguar range is the only area where abundant White-tailed Deer dominate the diet, for example 54 per cent of prey by biomass in the Chamela-Cuixmala Biosphere Reserve, south-western Mexico. Reptiles form a bigger part of Jaguar diet than that of any other large cat. In regions where medium-large reptiles are abundant and/or where Capybaras and peccaries are less available, reptiles may be the most important prey, for example Yacaré, Spectacled and Black Caimans in Amazon forest and some parts of the Pantanal; Jaguars take all life-stages of these three species, including large adults and eggs. Jaguars also prey frequently on at least another 14 reptile species, mostly large and medium-sized species. Large freshwater turtles (genus *Podocnemis*) and terrestrial tortoises (genus *Chelonoidis*) are important prey in Amazonian flooded forest and Brazilian Atlantic Forest (recorded in up to 25 and 20 per cent of scats, respectively). Jaguars are known to kill four species of marine turtles: Green, Leatherback, Hawksbill and Olive Ridley, chiefly when females are nesting in large aggregations on beaches. Green Turtles appear to be the most important prey species during the nesting season (June–October) in Tortuguero

National Park, Costa Rica; a minimum 672 adult Green Turtles (plus one Leatherback and three Hawksbills) were killed by Jaguars in 2005–2010. Snakes, including large anacondas and Boa Constrictors, are also killed by Jaguars although they do not contribute significantly to intake.

Depending on the relative availability of these two primary prey categories – medium-sized mammals and medium to large reptiles – Jaguars also eat a variety of smaller vertebrates: mammals such as primates, armadillos, tamanduas, agoutis and marsupials, and reptiles, chiefly iguanas and tegu lizards. During a period in which peccaries were rare in the Cockscomb Basin, Belize, Jaguars ate mainly Nine-banded Armadillos, Spotted Pacas and Red Brocket Deer. Following 20 years of protection at this site, armadillos are still the most important prey species (42 per cent of biomass consumed) followed by peccaries (15.6 per cent),

which are now more common. Similarly, in the Maya Biosphere Reserve, Guatemala, Nine-banded Armadillos and White-nosed Coatis are the most important prey species – 58.1 per cent of biomass for areas hunted by people and 46.3 per cent in better protected areas (where peccaries are more common); peccaries comprise 15.5 per cent of biomass and 27.2 per cent respectively. In Amazonian flooded forest, Jaguars feed mainly on caimans, Brown-throated Sloths and primates. The Giant Anteater is the most frequent prey item in *cerrado* savanna, central Brazil (Emas National Park) and in dry *caatinga* thorn-scrub of north-eastern Brazil. As the largest mammalian predator in Latin America, Jaguars kill a wide variety of other carnivores including records of Puma, Ocelot, Margay, Maned Wolf, Crab-eating Fox, Grey Fox, Tayra, skunks, Kinkajou, Cacomistle, olingos, raccoons and coatis. Killed carnivores are frequently abandoned without eating and rarely

A Jaguar narrowly misses a Capybara during a daylight hunt in the Brazilian Pantanal. In habitats with marked seasonality, Jaguars focus their hunting efforts along river systems during the dry season.

feature as important prey with the exception of procyonids; that is, White-nosed Coati, South American Coati and Crab-eating Raccoon, which comprise 5–21.5 per cent of biomass consumed in Mesoamerica and some areas in Brazilian Atlantic Forest and the Pantanal. Cannibalism occurs rarely including cases of cubs consumed by infanticidal males, and one incident in which two adult males killed and partially ate an adult female (southern Pantanal, Brazil). Jaguars incidentally eat birds, amphibians and fish. Rheas are important prey (13 per cent occurrence in scats) for Jaguars in the *cerrado* savannas, central Brazil. Jaguars in the Amazon are recorded killing freshwater dolphins as they fish in shallow water with records from Brazil and Colombia. Jaguars readily kill domestic livestock, chiefly cattle and occasionally horses. In

Unusual among felids, the Jaguar often kills large prey with a crushing bite to the skull or nape. Recent film of a Jaguar killing a large caiman using this technique shows the reptile paralysed instantly by the bite – a safe and efficient method for dealing with powerful, dangerous prey.

ranching habitats such as the Pantanal and Los Llanos, cattle form a major prey species, and are the most important prey for Jaguars in the southern Pantanal, Brazil (31.7 per cent of kills), and in north-eastern Sonora, Mexico (57.7 per cent of biomass consumed). Jaguars infrequently take domestic pigs from villages, and free-ranging dogs are killed, for example in forest on the Yucatán Peninsula near human communities. Jaguars almost never hunt people; most recorded attacks are caused by extreme provocation, such as during Jaguar hunts, and verified unprovoked attacks are extremely unusual.

Jaguars forage chiefly at dusk, night and early morning. Jaguars hunt primarily on the ground – they are poorly adapted to pursue prey arboreally – as well as in and around water sources. The Jaguar

is perhaps the most water-adapted felid and actively hunts in water. Jaguars pursue fleeing prey into water and launch spectacular leaping attacks from high riverbanks onto caimans and Capybaras in the water below. They also search for prey by passively floating downstream with the current to stealthily locate caimans and Capybaras resting on river beaches, and they readily launch attacks directly from the water. The Jaguar has a massively constructed skull equipping it with proportionally the strongest bite force of all the large cats. Although prey may be killed with a typically feline suffocating throat bite, Jaguars are unusual among cats for often killing large prey (including very large caimans and cattle) by a crushing bite to the skull, typically delivered at the rear of the braincase. This enormously powerful bite also allows Jaguars to open the carapace of large freshwater turtles and land tortoises. They kill large marine turtles by biting the vulnerable neck close to the skull.

Hunting success of Jaguars is unknown. They drag kills into dense cover, with anecdotal reports that dragging is more frequent and occurs over large distances in ranching landscapes where Jaguars are persecuted. A small female Jaguar with an estimated weight of 41kg dragged a 180kg heifer 200m through a densely forested ravine in Venezuela. Jaguars are not recorded covering kills. They return to large kills, which are consumed over a number of days. Jaguars readily scavenge, particularly from dead livestock that are the chief source of carrion in much of the range. Two different male Jaguars were recorded feeding on the carcass of a beached marine dolphin (species uncertain) on the northern Honduras coast.

Social and spatial behaviour

The socio-spatial ecology of the Jaguar is still relatively poorly known, though they clearly follow the fundamental felid pattern of being largely solitary and territorial. Both sexes maintain enduring territories in which the ranges of males are larger than those of females. Exclusive range use appears to be limited to small core areas; adults in some

populations have highly overlapping ranges, possibly due to marked seasonal changes in the distribution of water and hence prey. Pantanal females establish largely exclusive ranges during the wet season, but they overlap significantly in the dry season while males overlap extensively in both seasons. Similarly, Cockscomb Basin (Belize) males have highly overlapping ranges with some camera-trap locations visited by as many as five males per month.

As is typical for felids, male ranges overlap numerous female ranges, but high overlap among males means the reverse is also true for some populations. In one southern Pantanal study, one adult female's range was overlapped by at least three adult males, and her range was entirely encompassed by one of them. The following year, she overlapped with two males (one of them from the previous year and one new male); these same two males also overlapped a second adult female's range extensively.

Even with greatly overlapping ranges, it seems that adult Jaguars are no more sociable than other largely solitary cats. Adults engage in typically territorial behaviours, such as roaring and urine-marking, perhaps serving more to avoid encounters rather than to demarcate exclusivity. Of 11,787

A male Jaguar and the adult female Leatherback Turtle it has killed, with Black Vultures in attendance, Tortuguero National Park, Costa Rica. Despite their size, nesting marine turtles are powerless to defend themselves against Jaguars.

telemetry locations recorded over 3.5 years during the southern Pantanal research, there were 32 possible interactions between a female and a male jaguar, 21 possible encounters between two males, and only one possible encounter between two females. Two adult males that were clearly not siblings shared a feral pig carcass. Males are sometimes seen interacting amicably with females and their cubs, presumably when the male is the putative sire. Aggressive interactions between adults appear to be rare. Of 697 camera-trap images of 23 individual male jaguars in Cockscomb Basin, Belize, only three photographs showed serious flesh wounds or scars that could have been inflicted by another male. Nonetheless, there are occasional records of fatal fights; for example, an adult male was killed by another male presumably in a clash over territory in the southern Pantanal, Brazil.

Territory size varies with habitat quality and prey availability, from 30–47km² (females in moist savannas, Venezuela and Brazil) to 1291km² (male, dry savanna, Paraguay). Range size estimates are confounded by small sample sizes and few studies using GPS telemetry (which avoids the problem of underestimating range size that is common in conventional telemetry studies). Of studies with large samples and/or from GPS telemetry, average range sizes are available from wet Pantanal savanna woodland (females 57–69km²; males 140–170km²), Atlantic Forest, Brazil (females 92–212km²; males 280–299km²) and dry Gran Chaco savanna woodland, Paraguay (females 440km²; males 692km²). A lone male in arid lowland desert and pine-oak woodlands in Arizona used at least 1,359km² between 2004 and 2007 (calculated by camera-trapping and hence an underestimate). Two males in tropical forest in the southern Yucatán Peninsula, Mexico, had ranges of at least 1,000km². Home range sizes in seasonally inundated habitat (for example, the Pantanal) often contract in the wet season when flooded areas limit the space available to prey. The density of Jaguar populations increases along a cline loosely associated with precipitation. The lowest density populations are found in the arid regions of the range such as in Sonora, Mexico (1.05 Jaguars per 100km², Northern Jaguar Reserve, Mexico) and *caatinga* habitat, north-eastern Brazil (2.7 Jaguars per 100km², Parque Serra da Capivara, Brazil). The highest densities are recorded from very wet lowland forest (7.5–8.8 Jaguars per 100km², Chiquibul and Cockscomb, Belize) and wet savanna woodlands (6–7 Jaguars per 100km² and possibly as high as 11 Jaguars per 100km², Pantanal, Brazil).

Reproduction and demography

Reproductive patterns from the wild are very poorly known. Although it is common to find published references to a mating season in the Jaguar, reproduction is aseasonal. It is possible there are birth peaks in regions with marked seasonal

A female Jaguar and her two-month-old cubs cool off in Rio Cuiabá in the Brazilian Pantanal at the height of the dry season. There is almost no information on the survival and dispersal of wild Jaguar cubs.

differences, for example in Los Llanos and the Pantanal which have marked wet and dry seasons. Oestrus lasts 6–17 days and gestation lasts 91–111 days, averaging 101–105 days. Litter size is one to four cubs, averaging two (captivity). Weaning begins around 10 weeks and suckling typically ceases by four to five months. Cubs reach independence at 16–24 months. Dispersal is poorly known but it appears to be typically feline in which females settle close to their natal range while males disperse more widely. Inter-litter interval is unknown from the wild. Both sexes are sexually mature at 24–30 months; females first give birth at 3–3.5 years and can reproduce to 15 years (captivity). It is not known when wild males first breed but as for other large cats, it is likely to be no earlier than the age at which males are first able to become territorial; that is, no younger than three to four years old.

Mortality Mortality rates and factors in wild Jaguars are largely unknown. Adult jaguars have no natural predators and are killed principally by people and rarely by other Jaguars in territorial disputes. Adult Jaguars on the boundary of Calakmul Biosphere Reserve, Mexico, were close to human communities and were exposed to domestic carnivore diseases (feline heartworm and toxoplasmosis) while Jaguars deep inside the reserve were not; it is unknown if this led to mortalities. Predators of cubs are poorly known. Infanticide by male Jaguars is recorded, and there is one record of a female killing an unrelated cub at a cow carcass being used by the cub's mother and the infanticidal female in the Brazilian Pantanal.

Lifespan Poorly known from the wild, but is unlikely to exceed 15–16 years, and reaches 22 years in captivity.

STATUS AND THREATS

Jaguars have been extirpated from an estimated 49 per cent of their historical range and are extinct in El Salvador, the US and Uruguay. Despite this, the Jaguar's remaining range contains large areas of essentially continuous habitat, in part because the massive forested basins of South America have remained mostly inaccessible until recently. The most extensive Jaguar stronghold with a high probability of long-term persistence is the Amazon basin rainforest, and neighbouring areas of the Pantanal and Gran Chaco woodlands. Large fragments of tropical lowland forest in Central America are also considered to be strongholds, chiefly the Selva Maya of Mexico, Guatemala and Belize; the Río Plátano forest on the border of Honduras and Nicaragua; and a narrow strip of the Chocó-Darién moist forest from northern Honduras through Panama to northern Colombia. Most of the remaining range in Central America and Mexico is highly threatened. Similarly, populations at the periphery of remaining Jaguar range are regarded as severely threatened, particularly those in coastal dry forest in Venezuela; the Gran Sabana woodlands of Guyana, Venezuela and northern Brazil; the Atlantic Forest and *cerrado* of Brazil; and the Gran Chaco woodlands in northern Argentina. Populations in the inter-Andean valleys in Colombia represent the critical connection between Central America and South America, and are severely threatened.

Latin America has very high rates of habitat conversion for forestry, livestock and agriculture that directly threaten the Jaguar. This is combined with intense persecution from ranchers and pastoralists in livestock areas despite the fact that many cattle losses blamed on predation occur from other factors. People remove very large numbers of Jaguars in some ranching landscapes; for example, an estimated 185–240 large cats (Jaguars and Pumas) were killed between 2002 and 2004 in a 34,200km^2 ranching landscape in Alta Floresta, Brazil. Human hunting of prey is also likely to impact Jaguar populations and may be underestimated in areas of intact forest with human communities who rely heavily on the same prey species as large cats. There is also emerging evidence that the species is hunted to supply the Chinese medicinal trade, though this is poorly quantified. Sport-hunting is illegal in all range states, though illegal hunting is popular as recreation in some areas, for example the Pantanal and Los Llanos. There is no commercial hunting of Jaguars for their pelts since international markets were closed in the mid-1970s, though there is still widespread local demand for Jaguar paws, teeth and skins. CITES Appendix I. Red List: Near Threatened. Population trend: Decreasing.

Species overview

Wildcat
Felis silvestris

Chinese Mountain Cat
Felis bieti

Jungle Cat
Felis chaus

Sand Cat
Felis margarita

Black-footed Cat
Felis nigripes

Flat-headed Cat
Prionailurus planiceps

Pallas's Cat
Otocolobus manul

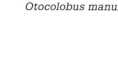

Rusty-spotted Cat
Prionailurus rubiginosus

Leopard Cat
Prionailurus bengalensis

Fishing Cat
Prionailurus viverrinus

Marbled Cat
Pardofelis marmorata

Bay Cat
Catopuma badia

Asian Golden Cat
Catopuma temminckii

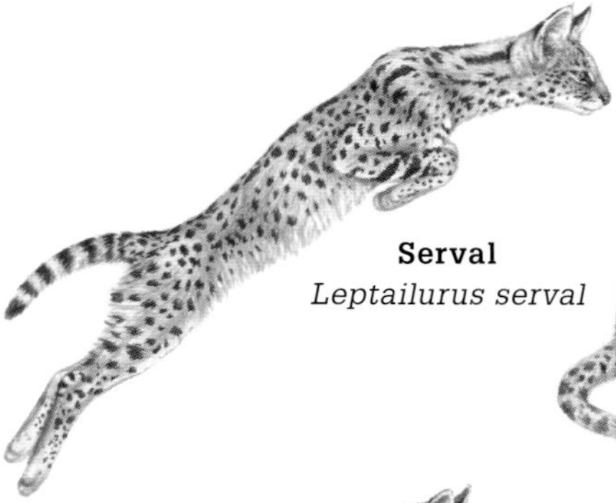

Serval
Leptailurus serval

Caracal
Caracal caracal

African Golden Cat
Caracal aurata

Geoffroy's Cat
Leopardus geoffroyi

Oncillas
Leopardus tigrinus;
Leopardus guttulus

Margay
Leopardus wiedii

Ocelot
Leopardus pardalis

Guiña *Leopardus guigna*

Colocolo
Leopardus colocolo

Andean Cat
Leopardus jacobita

Colocolo
Leopardus colocolo

Iberian Lynx
Lynx pardinus

Bobcat
Lynx rufus

Eurasian Lynx
Lynx lynx

Jaguarundi
Herpailurus yaguarondi

Canada Lynx
Lynx canadensis

Puma *Puma concolor*

Cheetah
Acinonyx jubatus

Snow Leopard
Panthera uncia

Clouded Leopards
Neofelis diardi;
Neofelis nebulosa

Lion
Panthera leo

Tiger   *Panthera tigris*

Leopard
Panthera pardus

Jaguar   *Panthera onca*

Conserving wild cats

Protected areas

Conserving cats starts with securing their habitat and prey in parks and reserves. Most of the world's Tigers now live in protected areas or in landscapes with a protected nucleus. If those reserves disappear, so too would the wild Tiger. The immense challenge facing most of the world's protected areas is inherent in the name – protection. As human populations continue to grow, the pressure for land and resources intensifies and so does illegal activity, clearing protected habitat and hunting wildlife. Even under protection, cats are not necessarily safe. The demand driven by traditional Asian medicinal beliefs for Tiger body parts (which have as much medicinal value as consuming a cow) is now so intense and the trade is so valuable that Tigers continue to be hunted in reserves across their range.

Yet protection clearly works. Tiger numbers have increased in India (Corbett National Park and the Western Ghats) and Nepal (Bardia and Chitwan National Parks) where governments, conservation NGOs and the donor community have committed the necessary resources to quell poaching and protect

forest. Many of the world's parks require a massive infusion of funding to achieve the same outcome. The Lion is now extinct in most of West Africa's protected areas because those nations are among the poorest on earth; they cannot afford to safeguard their parks and Lion populations gradually winked out, mainly from pervasive human hunting of their prey species for meat. The last 250 West African lions will persist only if the four parks in which they still remain are vigorously protected. Their case is particularly dire but it is a harbinger for many felid populations around the world. Without parks, very well insulated from the worst anthropogenic impacts, most cat species would decline and some would vanish entirely.

Resolving retaliatory killing of cats

Parks alone, especially poorly protected ones, do not guarantee the survival of many felid species. Human-modified landscapes dominate the globe and, despite their demanding ecological requirements, cats occupy many of them. Provided there is prey and

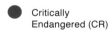

IUCN REDLIST KEY

● Critically Endangered (CR)
◉ Endangered (EN)
◉ Vulnerable (VU)
◉ Near Threatened (NT)
◉ Least Concern (LC)

ASSESSING STATUS

The IUCN Red List of Threatened Species (www.iucnredlist.org) evaluates species status based on multiple criteria including population size, degree of various threats, rate of population change and so on. From most to least threatened, the categories that apply to felids are Extinct (EX), Extinct in the Wild (EW), Critically Endangered (CR), Endangered (EN), Vulnerable (VU), Near Threatened (NT) and Least Concern (LC). All felids have been assessed at the species level and some regional populations or sub-species have been assessed separately, usually due to elevated concern, for example, the Lion in West Africa (Critically Endangered) compared to the Lion overall (Vulnerable). Until 2015 only one felid, the Iberian Lynx, was Critically Endangered at the species level, i.e. facing an extremely high risk of extinction in the wild. Following a massive conservation effort, it is now assessed as Endangered. Downlisting a species indicates a decreased likelihood of extinction, not that it is 'saved'. The Iberian Lynx is now more secure than when listed as CR in 2002 but it still faces serious threats and could decline again rapidly if conservation effort was relaxed. Downlisting is sometimes also the result of improved information – surveys indicating more populations than previously known, for example – rather than improved status as a result of conservation action. At the time of writing, a revision of Red List status is underway for all felids; the 2015 evaluation is provided where complete, otherwise it is from the previous assessment made in 2008 (indicated in the text).

A camera-trapped Tiger overlooks the town of Kotdwar in Uttarakhand state, India. Wild cats increasingly live uneasily at the interface of wilderness and encroaching humanity, where the challenges in conserving them are profound.

habitat, felids can live close to people, typically in significantly lower densities than under strict protection, but imperative for their conservation nonetheless. In contrast to the Tiger and Lion, many of the world's Cheetahs, Leopards, Snow Leopards and many other felids live outside formally protected areas. Anthropogenic landscapes also connect large protected tracts, allowing the movement of dispersers and their genetic contribution between core cat populations. However, wherever there are people – and especially their livestock – conflict is inevitable. People kill cats as a potential threat to our domestic animals and sometimes to people themselves, even when that threat is more perceived than real. Carnivores often kill very little livestock compared to other factors (although predation is occasionally devastating to individual herders) but pre-emptive and retaliatory killing of cats remains a global issue.

Saving cats in such landscapes relies on solutions that avoid conflict in the first place. Commercial sheep ranchers in Argentina use large livestock guarding dogs that mount an effective defense against Pumas, mitigating both predation on the herd and the traditional solution of employing a dedicated Puma bounty hunter to trap cats. In Snow Leopard range, conservationists vaccinate livestock against disease, a far greater source of mortality than carnivores. By reducing the losses to disease, herders can sell surplus stock to buy fodder for the lean winters which allows them to keep livestock in villages for an extra month or so when spring arrives. This keeps livestock away from the valley bottoms undergoing the spring flush of new grass, needed by wild ungulates that in turn support Snow Leopards. By the time livestock goes out grazing, Snow Leopards and their wild prey have dispersed and so too has the likelihood of conflict. Humans, livestock and cats can inhabit the same landscape, albeit often uneasily and with occasional losses on both sides, but it relies on investing in innovative and conscientious care of livestock. Unfortunately, many pastoralists around the world still choose the cheaper option of a bullet or poison.

Paying to stay; financial mechanisms to conserve cats

Whether inside protected areas or outside them, conservation is more likely to take root if cats have a dollar value to the communities that live nearby. Ecotourism is one obvious solution. Africa's great game reserves are famous for their big cats and are directly supported by the money spent by millions of tourists hoping to see them. India and Nepal's Tiger reserves are similarly reliant on tourism and, only in the last decade, opportunities have arisen to view Iberian Lynx, Jaguars, Pumas and Snow Leopards in the wild. Although indispensable, tourism revenue will only ever protect a relatively small fraction of wild cat distribution. West Africa's Lions declined precipitously in part because the region does not hold the tourism appeal enjoyed by the Serengeti or the Okavango Delta.

One option is to simply pay people to co-exist with cats. Most often, this takes the form of compensating people for the loss of livestock to carnivores in the hope it fosters tolerance. In reality, depredation is difficult to authenticate, carnivores are blamed for losses to other factors and falsifying claims inevitably creeps into the system. Being paid to put up with losses may even remove some incentive to reduce them, so the root cause of conflict goes unaddressed. Rather than paying for the negative value of living with cats, providing rewards or 'performance payments' is an intriguing alternative. Conservationists in northern Mexico pay ranchers for camera-trap photographs of Jaguars on private property. When the presence of Jaguars yields greater value than the cost of living with them (and it is crucial to adopt measures to reduce depredation alongside the payments), ranchers are less likely to kill cats. Similarly, where livestock is insured against losses to cats, herders have an incentive to better care for their herds. Just as with drivers whose vehicle insurance rates reflect the ability to avoid damage, farmers practising sound husbandry have fewer losses and lower premiums. Helping poor pastoralists contribute to the premium may be all the incentive they require to stop killing cats.

Captivity and captive breeding

Wild cats have been caged by people for millennia but does keeping cats in modern captivity contribute to their conservation? Captive populations potentially safeguard a reservoir of breeding individuals against the worst-case scenario of extinction in the wild. Saving the Iberian Lynx relied on a massive breeding program that now produces more Lynx kittens than can be accommodated in remaining wild habitat. It could not have succeeded without captive breeding although is the sole example of a captive felid population contributing directly to population recovery. A similar effort will soon release Persian Leopards in the Russian Caucasus; as with the Lynx, the offspring of captive adults will have opportunities to hunt in controlled conditions in the hope it better prepares them for survival in the wild. The vast majority of captive cats around the world will never be useful for this purpose.

More usefully, some zoos also contribute funding to *in situ* cat conservation (although globally it is an astonishingly tiny sum) and perhaps help foster a love for wild cats among urban human populations. The same cannot be said for circuses or Las Vegas stage shows keeping big cats for performances, which simply have no conservation value. Similarly, despite frequent assertions of conservation, private ownership and breeding of wild cats, especially in the United States, South Africa and a handful of other nations, plays next to no role in saving those species.

CONTROLLING TRADE

CITES (www.cites.org) is a treaty between 184 national governments to control international trade in live wildlife and their parts including furs, hunting trophies and souvenirs. The species covered by CITES are listed in three Appendices, according to the degree of protection they need. All wild cats are listed in Appendices I or II (the domestic cat is not classified). Appendix I covers species threatened with extinction, in which trade is permitted only in exceptional circumstances. Appendix II includes species not immediately threatened with extinction, but in which trade must be controlled in order to avoid utilization that may threaten their survival.

Further reading and resources

Major books and websites covering felids.

Anton, M., 2013. *Sabertooth*. Indiana University Press.

Bailey, T.N. 2005. *The African Leopard: Ecology and Behaviour of a Solitary Felid*. The Blackburn Press.

Caro, T.M., 1994. *Cheetahs of the Serengeti Plains: Group Living in an Asocial Species*. University of Chicago Press.

Divyabhanusinh, 2002. *The End of a Trail: the Cheetah in India*. Oxford University Press.

Divyabhanusinh, 2005. *The Story of Asia's Lions*. Marg Publications.

Gittleman, J.L., Funk, S.M., MacDonald, D.W., & Wayne, R.K. 2001. *Carnivore Conservation*, Cambridge University Press.

Hansen, K. 2006. *Bobcat: Master of Survival*. Oxford University Press USA.

Heptner, V. G. & Sludskii, A. A. 1992. *Mammals of the Soviet Union, Carnivora, Vol. II, Part 2. Hyaenas and Cats*, E. J. Brill.

Hoogesteijn, R. & Mondolfi, E. 1992. *The Jaguar*. Armitano Editores C.A.

Hornocker, M. & Negri, S. 2009. *Cougar Ecology and Conservation*, University Of Chicago Press.

Hunter, L. & Hinde, G. 2006. *The Cats of Africa: Behavior, Ecology and Conservation*. Johns Hopkins University Press/New Holland.

Hunter, L. & Hamman, D. 2003. *Cheetah*. Struik-New Holland.

Hunter, L. & Barrett, P. 2011. *Field Guide to Carnivores of the World*. New Holland/Princeton University Press.

Kingdon, J. & Hoffmann, M. (Eds). 2013. *The Mammals of Africa: Volume V: Carnivores, Pangolins, Equids and Rhinoceroses.* Bloomsbury Publishing.

Logan, K. A. & Sweanor, L. L. 2001. *Desert Puma: Evolutionary Ecology and Conservation of an Enduring Carnivore*. Island Press.

Macdonald, D. W. & Loveridge, A. J. (Eds). 2010. *Biology and Conservation of Wild Felids.*Oxford University Press.

McCarthy, T. M. & Mallon, D. (Eds). *in press. Snow Leopards of the World*. Elsevier.

Rabinowitz, A. 2014. *An Indomitable Beast: The Remarkable Journey of the Jaguar*. Island Press

Ruggiero, L.F., Aubry, K.B., Buskirk, S.W., Koehler, G.M., Krebs, C., McKelvey, K.S. & Squires, J.R. 2006. *Ecology and Conservation of Lynx in the United States*, University Press of Colorado.

Schaller, G.B. 1972. *The Serengeti Lion: A Study of Predator-Prey Relations*. University of Chicago Press.

Sanderson, J.G., & P. Watson. 2011. *Small Wild Cats: The Animal Answer Guide*. Johns Hopkins University Press.

Seidensticker, J. & Lumpkin,S. 2004 *Smithsonian Answer Book: Cats*. Smithsonian Books.

Sunquist, M. & Sunquist, F. 2002. *Wild Cats of the World*. University of Chicago Press.

Spalton, A., & al Hikmani, H.M. 2014. *The Arabian Leopards of Oman*. Stacey International.

Thapar, V. 2014. *Tiger Fire*. Aleph Book Company.

Tilson, R. & Nyhus, P. 2010. *Tigers of the World: The Science, Politics and Conservation of Panthera tigris*. Academic Press

Turner, A. & Anton, M. 1997. T*he Big Cats and Their Fossil Relatives: an Illustrated Guide to their Evolution and Natural History*. Columbia University Press.

Wilson, D.E. & Mittermeier, R.A. (Eds). 2009. *Handbook of Mammals of the World, Vol. 1: Carnivores*, Lynx Edicions.

Websites

Carnivore Ecology & Conservation An excellent compendium of carnivore news and knowledge compiled by carnivore biologist Guillaume Chapron. www.carnivoreconservation.org

IUCN/SSC Cat Specialist Group. Information on the world's cats including the biannual journal *Cat News* and an outstanding online library of thousands of scientific papers and reports. www.catsg.org

Panthera. The largest conservation NGO working exclusively on conserving the world's wild cats. www.panthera.org

Conversion Table

Throughout this book measurements, weights and areas have been provided using the metric system, however, those more used to the Imperial system may find the following table useful.

Metric to Imperial Conversion Chart					
Metric units	**Imperial units**		**Metric units**	**Imperial units**	
Length			**Area**		
1cm	0.39in (⅜in)		1m²	1.20yd²	
5cm	1.9in (1¹⁵⁄₁₆in)		5m²	5.97yd²	
10cm	3.94in (3¹⁵⁄₁₆in)		10m²	11.96yd²	
			25m²	29.90yd²	
1m	3.28ft				
3m	9.84ft		100km²	39 sq mi	
5m	16.40ft		1000km²	390 sq mi	
10m	32.81ft				
15m	49.21ft		**Weight**		
25m	82.02ft		10g	0.35oz (⅓oz)	
50m	164ft		50g	1.76oz (1¹⁹⁄₂₅oz)	
100m	328ft		1kg	2.2lb	
1000m	3,280ft		3kg	6.6lb	
			5kg	11lb	
1km	0.621mi		10kg	22lb	
10km	6.21mi		50kg	110.2lb	
100km	62.14mi		100kg	220.5lb	

For temperature from celsius multiply by 9, divide by 5, then add 32 for fahrenheit, 0°C = 32°F

Acknowledgements

I am deeply grateful to the following colleagues who reviewed sections of the text; George Amato, Mauricio Anton, Vidya Athreya, Laila Bahaa-el-din, Christine and Urs Breitenmoser, Arturo Caso, Passanan Cutter, Pete Cutter, Tadeu Gomes de Oliveira, Will Duckworth, Sarah Durant, Mark Elbroch, Paul Funston, John Goodrich, Sanjay Gubbi, Andy Hearn, Philipp Henschel, Marna Herbst, Rafael Hoogesteijn, Örjan Johansson, Roland Kays, Marcella Kelly, Andrew Kitchener, John Laundre, Mauro Lucherini, Quinton Martins, Jennifer McCarthy, Tom McCarthy, David Mills, Gus Mills, Dale Miquelle, Shomita Mukherjee, Constanza Napolitano, Aletris Neils, John Newby, Andres Novaro, Kunel Patel, Esteban Payan, Howard Quigley, Seth Riley, Joanna Ross, Steve Ross, Jim Sanderson, Elke Schüttler, Tanya Shenk, Alex Sliwa, Chris and Tilde Stuart, Lilian Villalba, Tim Wacher, Susan Walker, Byron Weckworth, Andreas Wilting, and Guillermo López Zamora.

My thanks go to many people who donated images for this book; Laila Bahaa-el-din, Paulo Boute, Raghu Chundawat, Nick Garbutt/nickgarbutt.com, Laurent Geslin/laurent-geslin.com, Melvin Gumal, Urs Hauenstein, Andy Hearn/Jo Ross, Philipp Henschel, Chitral Jayatilake,

Paul Jones, Leo Keedy, Emmanuel Keller, Sebastian Kennerknecht/pumapix.com, Hyuntae Kim, Patrick Meier/mywilderness.net, Manuel Moral, Jerry Laker/Fauna Australis, Rodrigo Moraga/natphoto.cl, Antonio Nuñez-Lemos/ anunezlemos.com, Pete Oxford/peteoxford.com, Parinya Padungtin, Hardik Pala, David Palacios, Bivash Pandav, Frank Reynier, Barry Rowan/barryrowan.com, Alfonso Tapia Sáez, Santosh Saligram, Octavio Salles/octaviosalles.com.br, Tashi Sangbo, Alex Sliwa, Christian Sperka/sperka.biz, Chris and Tilde Stuart, Nina Sunden, Mauro Tammone, Gavin Tonkinson/gavintonkinson.com, James Tyrrell, Yvonne van der Mey, Rodrigo Villalobos Aguirre, Helen Young, Tim Wacher, Larry Wan/wanconservancy.com.

At Panthera, I am grateful to Danielle Garbouchian for help with the sabertooth chart; to Michael Levin and Becca Marcus for helping compile images; and especially to Lisanne Petracca for the range maps; and to Andrew Williams for improving the quality of camera-trap photos. Special thanks to Peter Gerngross for his excellent map of Leopard range, and to the Range Wide Conservation Program for Cheetah and African Wild Dogs for updated range data on Cheetahs.

Image credits

Bloomsbury Publishing would like to thank the following for providing photographs and for permission to produce copyright material. While every effort has been made to trace and acknowledge all copyright holders, we would like to apologise for any errors or omissions and invite readers to inform us so that corrections can be made in any future editions of the book.

Maps by Lisanne Petracca/Panthera

All colour and black and white illustrations by Priscilla Barrett, except p11 *Smilodon* by Luke Hunter

Charts p7, 8, 237 by Julie Dando, Fluke Art

Photographs

Key t=top; tl=top left; right; tr=top right; b=bottom; bl=bottom left; br=bottom right

Photo libraries – FLPA: Frank Lane Photography Agency; NPL: Nature Picture Library; SH: Shutterstock

Throughout the book we have tried to show examples of wild cats in the wild. However, in the absence of such images, photographs of cats taken in captivity have been indicated in the caption with a 'c'.

Front cover: t/c SH; **Back cover:** t/c FLPA; **half-title:** FLPA; **contents:** Frans Lanting/FLPA; **11** James Tyrrell; **12** Sebastian Kennerknecht; **15** Sebastian Kennerknecht; **16–17** FLPA; **19** Tashi Sangbo; **22** Bernard Castelein/NPL, **23** Michael Breuer/Biosphoto/FLPA; **24** Alain Mafart-Renodier/FLPA; **25** Tim Wacher; **26** Ann & Steve Toon/NPL; **30, 31, 32, 33** Alex Sliwa; **35** Gerard Lacz/ FLPA; **36** Terry Whittaker/FLPA; **37** Gerard Lacz/FLPA; **39, 40, 41** HardkiPala; **43** Emmanuel Keller; **44** Christian Sperkka; **47** Hyuntae Kim; **48** Nick Garbutt; **49** Gerard Lacz/FLPA; **50** Gerard Lacz/FLPA; **52** Nick Garbutt/NPL; **55** Santosh Saligram; **56** Terry Whittaker/FLPA; **57** David Hosking/FLPA; **59** Nick Garbutt; **60** Parinya Padungtin; **61** Terry Whittaker/FLPA; **63** Sebastian Kennerknecht; **66** tr Andrew Hearn/Joanna Ross, b Sebastian Kennerknecht; **69** t Terry Whittaker/ FLPA, b JNPC/DWNP/Panthera/WCS Malaysia; **70** Gerard Lacz/ FLPA; **73** Nina Sunden; **74** bl Laila Bahaa-el-din, br Gavin Tonkinson; **76** Mitsuaki Iwago/FLPA; **77** Patrick Kientz/FLPA; **78** Denis-Huot/ NPL; **80** t C&M Stuart Controlled, b Yva Momatiuk & John Eastcott/ FLPA; **81** Luke Hunter; **82** Sebastian Kennerknecht; **83** Yvonne van der Mey; **84** Anup Shah/NPL; **87** t Laila Bahaa-el-din, b Sebastian Kennerknecht; **89** t Sebastian Kennerknecht, b Laila Bahaa-el-din/ Panthera; **90** Philipp Henschel; **92** tr Paul Jones, bl Gabriel Rojo/ NPL; **94** Rodrigo Villalobos; **97** Alex Sliwa; **98** Tadeu de Oliveira; **101** Frank Reynier; **102** Terry Whittaker/FLPA; **105** Patrick Meier, **106** Patrick Meier; **107** b Patrick Meier, br Patrick Meier; **108** Larry Wan; **109** Roland Seitre/NPL; **110** Patrick Meier; **113** t Rodrigo Moraga b Alex Sliwa; **114** Fauna Australis/Jerry Laker; **115** Mauro Tammone; **118** tr Luciano Candisani/FLPA; b Sebastian Kennerknect; **120** Alfonso Tapia Saez, **121** Pablo Dolsan, NIS/FLPA; **123** AGA/Rodrigo Villalobos; **124, 125, 126** Antonio Nuñez-Lemos, Antonio Nuñez-Lemos; **128** Bernd Rohrschneider/FLPA; **129** Willi Rolfes/FLPA; **130** Laurent Geslin; **131** Jules Cox/FLPA; **132** Laurent Geslin; **135** David Palacios; **136** Manuel Moral; **137** Wild Wonders of Europe/Pete Oxford/NPL; **138** David Palacios; **141, 142, 143, 144** Barry Rowan; **147** t Michael Quinton/Minden Pictures/FLPA, b Tim Fitzharris/Minden Pictures/FLPA; **148** Michael Quinton/Minden Pictures/FLPA; **149** Mark Newman/FLPA; **150** Michael Quinton/Minden Pictures/FLPA; **151** Chris and Tilde Stuart/FLPA; **153** Urs Hauenstein; **154** Panthera/ SBBD/IDB/ICE; **155** Panthera Colombia; **156** Gerard Lacz/FLPA; **158** Francois Savigny/NPL; **159** Ignacio Yufera/FLPA; **160** Rodgrio Moraga; **161** Sumio Harada/Minden Pictures/FLPA; **162** Sebastian Kennerknecht; **163** Octavio Salles; **164** Thomas Mangelsen/FLPA; **166** Jurgen & Christine Sohns/FLPA; **168** t Frans Lanting/FLPA, b I.R.I DoE/CACP/WCS; **170** Imagebroker/J.rgen Lindenburger / FLPA; **171** Luke Hunter; **172** Nick Garbutt; **173** Winfried Wisniewski/ FLPA; **174** t Frans Lanting/FLPA, b Michel & Christine Denis-Huot/ Biosphoto/FLPA; **175** Suzi Eszterhas/FLPA; **176** Laurent Geslin/NPL; **178** Raghu Chundawat; **179** Jeff Wilson/NPL; **182** Hiroya Minakuchi/ FLPA; **184** Sebastian Kennerknecht; **185** Christian Sperka; **186** Sebastian Kennerknecht; **188** Christian Sperka; **191** Nick Garbutt; **192** Hiroya Minakuchi/FLPA; **193** Vladimir Medvedev/NPL; **194** Andrew Parkinson/NPL; **195** Gerard Lacz/FLPA; **197** Imagebroker. net/FLPA; **198** Andy Rouse/NPL; **200** ImageBroker/FLPA; **201** Imagebroker, Fabian von Poser/Imagebroker/FLPA; **202** Christian Sperka; **203** Brendon Cremer/Minden Pictures/FLPA; **204** Patrick Meier; **205** Brendon Cremer/Minden Pictures/FLPA; **206** Christian Sperka; **207** Patrick Meier; **208** Richard Du Toit/Minden Pictures/ FLPA; **211** t Michael Durham/Minden Pictures/FLPA, b Parinya Padungtin; **213** Chitral Jayatilake; **214** Chitral Jayatilake; **215, 216** Nick Garbutt; **217** Helen Young; **218** Pete Oxford/Minden Pictures/ FLPA; **220** Patrick Meier; **221** Pete Oxford; **222** Nick Garbutt; **223** Patrick Meier; **224** Suzi Eszterhas/FLPA; **225** Leo Keedy/GVI; **226** Paulo Boute; **233** Bivash Pandav/OETI/Panthera.

Index